Florida Weather and Climate

UNIVERSITY PRESS OF FLORIDA

Florida A&M University, Tallahassee
Florida Atlantic University, Boca Raton
Florida Gulf Coast University, Ft. Myers
Florida International University, Miami
Florida State University, Tallahassee
New College of Florida, Sarasota
University of Central Florida, Orlando
University of Florida, Gainesville
University of North Florida, Jacksonville
University of South Florida, Tampa
University of West Florida, Pensacola

Florida Weather and Climate

MORE THAN JUST SUNSHINE

JENNIFER M. COLLINS, ROBERT V. ROHLI,
AND CHARLES H. PAXTON

University Press of Florida
Gainesville · Tallahassee · Tampa · Boca Raton
Pensacola · Orlando · Miami · Jacksonville · Ft. Myers · Sarasota

First cloth printing, 2017
First paperback printing, 2019

26 25 24 23 22   6 5 4 3 2

Library of Congress Cataloging-in-Publication Data
Names: Collins, Jennifer M., author. | Rohli, Robert V., author. | Paxton,
Charles H., author.
Title: Florida weather and climate : more than just sunshine / Jennifer M.
Collins, Robert V. Rohli, and Charles H. Paxton.
Description: Gainesville : University Press of Florida, 2017. | Includes
bibliographical references and index.
Identifiers: LCCN 2017010457 | ISBN 9780813054445 (cloth)
ISBN 9780813064284 (pbk.)
Subjects: LCSH: Florida—Climate. | Climatology. | Storms—Florida.
Classification: LCC QC984.F6 C65 2017 | DDC 551.65759—dc23
LC record available at https://lccn.loc.gov/2017010457

The University Press of Florida is the scholarly publishing agency for the State University
System of Florida, comprising Florida A&M University, Florida Atlantic University, Florida
Gulf Coast University, Florida International University, Florida State University, New College
of Florida, University of Central Florida, University of Florida, University of North Florida,
University of South Florida, and University of West Florida.

University Press of Florida
2046 NE Waldo Road
Suite 2100
Gainesville, FL 32609
http://upress.ufl.edu

Contents

Figures

Maps

Tables

Acknowledgments

The authors would like to thank Michelle Saunders, Amy Hyler, Kevin Ash, Jason Krzyzanowski, Anita Marshall, Melissa Metzger, and Amanda Gillum for their contributions to the figures in this book. We thank Heather Key for coordinating the team of students who helped generate the figures. Her efforts were instrumental in the completion of this book. We thank Leilani Paxton for her assistance in seeking permissions. We thank Ed Frederick for giving us permission to use an excerpt from his book *Ten Seconds inside a Tornado*. We appreciate the time Dr. Bryon Middlekauff, Wendy Collins, and Amy Polen took to carefully review earlier drafts of the chapters and the other experts who reviewed the book for the University Press of Florida. Finally, we deeply appreciate the productive collaboration with the professionals at the University Press of Florida.

1

The Sunshine State

Visitors driving into Florida pass signs that offer a greeting and mention the weather they may expect: "Welcome to Florida—The Sunshine State." It is the sunshine that lures so many people to Florida. Florida has been officially known as the Sunshine State since 1970, but other nicknames include the Peninsula State, the Everglades State, the Alligator State, and the Orange State. Florida is the land of sunshine, beaches, theme parks, wildlife, and much more.

Florida's warm, sunny weather has created a population of tourists, retirees, refugees from the cold North, and visitors from overseas, but Florida's weather isn't always warm and sunny. Every Floridian knows to expect rumbling thunderstorms arising from the hot and humid air lifted by sea breezes that sometimes turn violent during summer afternoons. These storms, which are sometimes deadly, throw out bolts of lightning like the Greek god Thor, striking whomever may be in their path. At times the weather is hot and dry for months, as Floridians all over the state saw in 1998 when the parched land ignited into wildfires. Thick smoke from fires at night clings close to the ground and mixes with fog, creating a deadly mix when it drifts over roadways. During the winter, Arctic air masses originating in Siberia migrate across the dark and frozen northlands to Alaska and Canada and plunge southward, bringing a frigid chill to Florida. Snow is not on the mind of most Floridians unless they are headed to the ski slopes or hear news of a blizzard up north. On occasion, though, northern Florida receives a white

Figure 1.1. The roadway greeting sign upon arrival in Florida. Credit: Chris Allen.

coating of measurable snow that sometimes dips even farther south to Tampa and Miami, as happened in January 1977. As nearshore ocean and gulf waters of central and north Florida chill repeatedly during cold outbreaks, warm, moist air moving from over the warm Florida Current and the Gulf Stream create huge sheets of fog and drizzle. Through the years, Florida residents have seen their fair share of violent and destructive weather in the form of major hurricanes, powerful tornadoes, and grapefruit-sized hail. Florida has gained many nicknames related to weather events. After the 2004 hurricane season, when Florida was slammed by four hurricanes, several nicknames were tossed around including the Hurricane State, the Plywood State, and the Blue Tarp State. At other times Florida has been known as the Lightning State, the Sinkhole State (particularly during drought times when sinkholes are more likely to form), the Tornado State, and the Frozen Orange State (during Arctic blasts of icy air).

Florida by the Numbers

Florida, the southeasternmost state, stretches almost 800 road miles from Pensacola in the western Panhandle through Orlando in the middle of the peninsula to Key West at the southernmost part of the Florida Keys. This distance is about the same as the width of Texas, but Florida's land area is narrow. Its total area, over 54,000 square miles ranks 22nd among the states. The peninsula ranges from 100 to 140 miles in width and almost 447 miles from north Florida to the tip of the Florida Keys.

Florida has thousands of freshwater lakes. The 700-square-mile Lake Okeechobee in the middle of the peninsula's southern end is the largest. Freshwater sources within the state encompass 4,300 square miles. By air, one can see that Florida is dotted with lakes; over 7,700 are larger than 10 acres. More than 11,000 miles of rivers, streams, and canals drain the rain that falls in Florida. The longest river in Florida, the St. Johns, is also the longest north-flowing river (at 273 miles) in the United States east of the Rocky Mountains. It runs along the eastern side of Florida from the St. Johns Marsh near Vero Beach in Indian River County to Jacksonville, where it bends to the east and meets the Atlantic Ocean.

Florida has about 4,500 islands larger than 10 acres; only the state of Alaska has more. Bordered by the Atlantic Ocean to the east and the Gulf of Mexico to the west, Florida has 1,200 miles of coastline, more than California's 840 miles. Florida is also known for its flat landscape, much of which is barely above sea level. The highest point in the state, at Britton Hill near the Alabama border, not far from the Panhandle town of Paxton in Walton County, is only 345 feet above sea level. This is the lowest high point among the 50 states.

Most of Florida's population lives near the Atlantic Ocean or the Gulf of Mexico. About 75 percent of the population lives in coastal counties. In 1845, when Florida became the 27th state, it had a population of only 66,500. According to the U.S. Census Bureau (2015), Florida's population estimate for 2014 was over 20 million people, making it the third most populated state; only California (38 million) and Texas (26 million) have higher populations. These residents live in over 9 million housing units in Florida's 67 counties. Twenty-one percent of these residents were under the age of 18 in 2014, and 18 percent were over the age of 64. Florida's

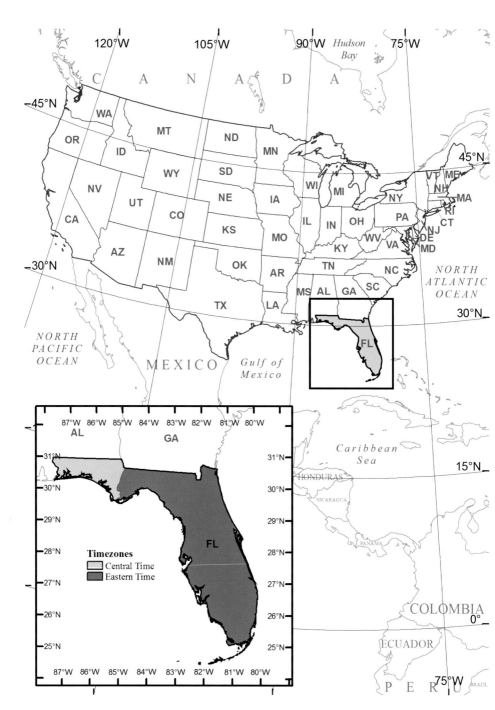

Map 1.1. The geographic grid and Florida time zones.

population density of 351 people per square mile ranks eighth among the states.

North Florida, which shares borders with Georgia and Alabama, is wide enough to be located in two time zones: Eastern and Central (map 1.1). The northernmost "point" of Florida actually extends in a west-east line over 100 miles along Florida's northern border with Alabama. The northernmost part of Florida is farther south than any part of the south-western states of New Mexico, Arizona, or California.

From 1938 to 1977, Florida automobile license plates indicated the 1938 population ranking of the county in which the vehicle was regis-tered. For example, Dade County (Miami) which was the most populated county in 1938, was number 1, Duval County (Jacksonville) number 2, Hillsborough County (Tampa) 3, Pinellas County (St. Petersburg) 4, and Polk County (Lakeland-Winter Haven) 5. Currently the top five county populations (with the largest city listed) have over one million residents and the top three reflect the population explosion that occurred along the lower southeast coast: 1) Miami-Dade (2.6 million; Miami), 2) Bro-ward (1.8 million; Fort Lauderdale), 3) Palm Beach (1.4 million; West Palm Beach), 4) Hillsborough (1.3 million; Tampa), and 5) Orange (1.2 million; Orlando). The city with the greatest areal size, 750 square miles, also has the greatest population, Jacksonville (821,000). Next in line by population, according to the 2010 U.S. Census, are Miami (399,000), Tampa (336,000), St. Petersburg (245,000), and Orlando (238,000).

Florida's warm climate is also a hospitable host to many nonhuman heat-loving creatures. Alligators, one of the more notably dangerous in-habitants, move silently through the state's lakes and rivers. They can reach 13 feet long and weigh up to 1,000 pounds. Nearly every year three to six people are bitten by alligators without provocation, and every year or two someone is killed by an alligator. Some Florida mammals that are considered dangerous are rarely seen. However, the numbers of these hunters are in decline. Only about 3,000 Florida black bears still exist and the Florida panther, the state animal, is nearly extinct; only about 180 exist today. They feed partly on the many deer, opossum, and ar-mored armadillos that roam the state. The climate is also favorable for six-legged insect annoyances, which include mosquitoes, fire ants, sev-eral varieties of cockroaches, and the notorious love bugs that splatter on automobiles during the months of May and September. As flood waters rise, many of the insects and spiders that live near and under the ground appear. Entire colonies of fire ants evacuate their tunnels and create a floating raft with their larvae on top when flooding occurs.

The Geographic Grid and Florida

We know that Florida is in the southeastern corner of the continental United States, between the Gulf of Mexico and the Atlantic Ocean, but how can we describe Florida's location more precisely? We often use a geographic grid of latitude (east-west parallels) and longitude (north-south meridians) to describe the location of any place on Earth (map 1.1). For example, the landfall location of Hurricane Charley, which landed in southwest Florida in 2004, was near Port Charlotte, Florida. This location can be stated several ways: in decimal form (i.e., 27.0°N, 82.1°W), in degrees and minutes (i.e., 27°00'N, 82°06'W), or with a negative sign that indicates longitudes west of the prime meridian (i.e., 27°00', -82°06').

Map 1.1 shows Florida along with several key lines of latitude and longitude. The easternmost part of Florida is near West Palm Beach, and the westernmost point is the Perdido River in Escambia County. The easternmost Florida longitudes are still west of major Atlantic Coast cities such as Boston, New York, Philadelphia, and Washington, D.C. Most people are surprised to learn that even Pittsburgh, in western Pennsylvania, is located east of the easternmost point of Florida at 80°W. In addition, the westernmost point in Florida is located almost as far west as Chicago. Florida extends farther west in the United States than most people realize. The Florida Keys are the southernmost part of the state, and Ballast Key, located at 24°31'N latitude, is the southernmost island.

The Sun and Seasons

Due to the earth's tilt and its revolution around the sun, the sun is directly overhead at noon at the Tropic of Cancer (23.5°N latitude) on June 21 (summer solstice) and at the equator (0°) on September 22 (equinox). It is directly over the Tropic of Capricorn (23.5°S latitude) on December

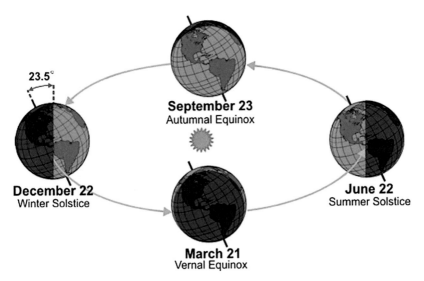

23.5°

September 23
Autumnal Equinox

December 22
Winter Solstice

June 22
Summer Solstice

March 21
Vernal Equinox

Figure 1.2. The seasons as Earth revolves around the sun. Source: NOAA (n.d.).

22 (winter solstice), then back over the equator by March 21 (equinox). These dates can vary by a day or two in some years. All of Florida is located to the north of the Tropic of Cancer, the northernmost latitude that ever experiences the overhead rays of the sun. Places north of the Tropic of Cancer never experience the sun directly overhead. In the summer, Florida is close to the intense overhead rays of the sun, but in the winter it is much farther from the latitude where the sun shines directly overhead, as is the case for the rest of the continental United States.

Another factor related to latitude that affects the amount of energy from the sun is the length of daylight hours. Because of the geometry of the earth's tilt as it orbits the sun, locations north of the Tropic of Cancer and south of the Tropic of Capricorn have fewer hours of daylight in winter than they do in summer and the least sunlight at the time of the winter solstice. The less direct rays of sunshine (i.e., at angles farther from being directly overhead) and the fewer hours of daylight mean that locations farther from the tropics are colder than places closer to the equator.

The Sun's Energy Keeps Florida Warm

The Sunshine State isn't always sunny; instead, Florida's weather and climate are interesting and active. Energy from the sun drives all of Earth's weather and climate. It takes about 8 minutes and 18 seconds for the sun's energy to radiate through space and into our atmosphere. The waves of electromagnetic energy from the sun travel at a variety of wavelengths. The length of a wave determines how it behaves in the atmosphere. For example, ultraviolet wavelengths are more easily absorbed in the atmosphere before they reach the earth's surface. Only a small part of the sun's radiant energy is in the ultraviolet part of the solar wavelength spectrum. This part of the spectrum is needed for the body to produce vitamin D but is also responsible for sunburns. Energy at other wavelengths is detected by our eyes as visible radiation in the form of different colors and accounts for about half of the sun's energy. Plants use the energy in the visible spectrum in photosynthesis. Although the sun radiates energy across a wide range of wavelengths, essentially all of this energy is at short wavelengths of less than about 4.5 millionths of a meter. Much of the shortwave energy from the sun that the earth absorbs is re-radiated as heat energy at longer wavelengths. Amazingly, only half of one billionth of the sun's radiation reaches the top of the earth's atmosphere. Then, because of clouds, gases, and tiny solid particles called particulates or aerosols that are suspended above the earth, only about half of that energy penetrates all the way to the earth's surface. Yet this is enough energy to sustain the planet, and it's more than enough to keep Florida warm most of the time. However, Florida isn't always warm and it isn't always sunny, despite what the license plates say, because the energy availability from the sun at a particular place is always changing, from day to night, from summer to winter, and at even longer time scales.

The amount of the sun's energy that reaches Florida varies from winter to summer, which in turn affects the temperature. Other factors can modify the amount of solar energy that reaches the surface of the earth. Dense cloud decks can block much of the incoming shortwave radiation during the day, keeping temperatures cooler, but preventing the earth's re-radiated longwave energy (heat) from escaping to space, keeping nighttime temperatures warmer. Spotty clouds can create changes in the distribution of solar energy and differences in localized temperatures. At longer time scales, Earth's long orbital cycles around the sun lead to changes in climate, from an ice-covered earth to globally warm periods.

At the same moment that a Floridian watches a beautiful sunset over the Gulf Coast and detects solar radiation decreasing, someone else half a world away is seeing the sun rise and feeling the effects of increasing solar energy. Someone else, somewhere else, at the same time, is experiencing an afternoon with an overcast sky that reduces the amount of energy that location is receiving. At yet another place, at the same instant, someone at a mountainous location is feeling the effects of solar energy that reaches the surface mostly unobstructed by clouds, atmospheric gases, or aerosols. Another person in the tropics, where the sun passes nearly overhead at noon, is experiencing intense rays of sunshine. Close to the North Pole, where the sun is at a much lower angle and doesn't come close to being overhead, even at noon, the sun's rays can be obstructed more easily as they pass through the atmosphere at a more oblique angle. And, of course, half of the earth is experiencing the darkness of night, facing away from the sun and receiving no solar radiation at all. The net effect of solar radiation for the earth is that the amount of solar energy that the earth and its atmosphere intercept stays nearly constant on a day-to-day basis.

To make things even more interesting, the sun's energy doesn't just heat the surface or the gases and solid aerosols that make up the atmosphere. The sun's differential heating also drives motion in the air—the winds—which in turn steer the ocean currents. Computer models tell us that if the sun suddenly burned out, atmospheric circulation would grind to a halt within a few weeks (in addition to all of the other problems this would create).

The Sun's Effect on Water

The heat from solar energy also evaporates water and melts ice. Water is a peculiar thing—it is the only common substance that changes phase between solid, liquid, and gas at the normal range of temperatures experienced on Earth. However, the changes in its state from solid ice to liquid water or from liquid water to water vapor (evaporation) requires an input of energy (figure 1.3). That energy input comes from the sun.

A bucket of water in Florida will evaporate faster in July than in January, and a block of ice will melt faster in July than January. The energy from the sun that went into evaporating the bucket of water or melting the ice could not be used to heat the surface of the earth or the atmosphere because it was used at the molecular level in the water or ice to cause the phase change. Evaporation and melting can be thought of as cooling the environment because the sun's energy is used at the molecular level to trigger the phase changes instead of being used to warm up the air and the earth's surface. For the same reason, when you perspire, it is the evaporation of that perspiration that makes you feel cooler. Even if the temperature is 90°F, you may feel cold when you get out of a swimming pool. This is because the water is suddenly evaporating from your skin, leaving your skin cooler.

When water changes phase in the other direction—from water vapor to liquid water (condensation)—the same amount of solar energy is released that was used to cause the phase change of a given mass of water in evaporation. Similarly, when water changes phase from liquid to solid (freezing), the same amount of solar energy that causes an equal amount of ice to melt is released in freezing.

The energy that is absorbed in evaporation or melting and released in condensation or freezing of water is called latent energy. When a cloud forms from condensation, the latent energy released by an unimaginably large number of transformations of water molecules triggers an equally unimaginably large number of molecular-scale perturbations that add up to enough energy to power winds, such as the circulations around powerful thunderstorms, tornadoes, and hurricanes. Without localized condensation and freezing when clouds form, the energy wouldn't be available to generate such strong winds in storms. Fair-weather days tend to have light winds because latent energy isn't as readily available. However, the latent energy available in even a small thunderstorm is tremendous—enough to power the entire United States for several minutes, if only someone could figure out how to harness such energy efficiently. The energy in a large thunderstorm could power the entire United States for several hours and the energy in a hurricane could do so for several days!

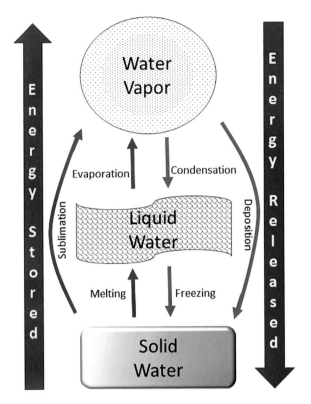

Figure 1.3. Phase changes of water.

The sun's energy drives not only the temperatures but also the weather we experience. The availability of energy from the sun at a local place changes every minute of the day, as the sun climbs toward noon or falls toward sunset or as a cloud moves between the sun and the observer. From this perspective, it's easy to see that the weather is always changing. Florida often has a ready supply of energy from the sun and water waiting to change phase, so the weather changes quickly, and all too often it changes powerfully too. Weather is ultimately about how the sun's energy and the presence of water team up to produce an ever-changing, awe-inspiring, and sometimes scary arrangement of sunshine, clouds, and storms—sometimes all within a few minutes and a few miles of each other!

2

Florida's Climate Types and Temperature

What comes to mind when Florida's climate is mentioned? Bright, sunny, and cool weather may come to mind for someone who has visited Florida in January; someone who has visited during the summer may think of oppressively hot and humid conditions. However, the climate of the state isn't composed of conditions in a single moment. It is the long-term pattern of weather conditions of a location. While it is tempting to think of climate as the average weather over time, the long-term pattern includes far more than just averages. The frequency of extreme weather conditions is also a part of a location's climate. After all, two different places could have the same average temperature but one could have more consistent temperatures with fewer extremes, as we might see at a location near the ocean where the water moderates the temperature, and the other location could have higher high temperatures and lower low temperatures, as we might see in a drier setting farther from the ocean. Florida has two generalized types of climates: tropical and subtropical.

Tropical Florida

The term "tropical" generally means that the temperature throughout the year is nearly constant, that temperatures are almost always warm. Technically, places between the Tropic of Cancer (23.5°N) and the Tropic

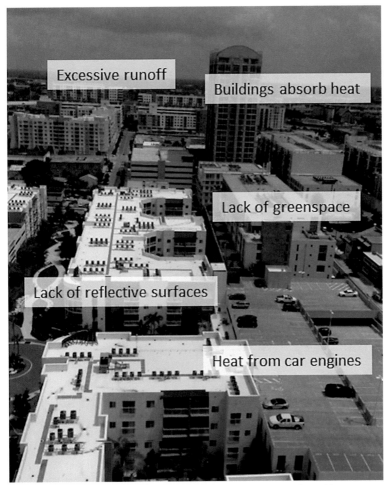

Map 2.1. Climate types and locations where Florida's record high and low temperatures were observed.

of Capricorn (23.5°S) are called the tropics. However, tropical climates can extend beyond these latitudes, as we see in Florida (map 2.1), as long as the average temperature in every month of the year exceeds 64°F (18°C) and conditions are not arid. In tropical climates, the afternoon high temperatures will be in the 70s or 80s Fahrenheit throughout the year. Even though south Florida, including the Everglades, falls outside the tropics as defined by geography, it has a tropical climate because the average monthly temperature exceeds 64°F (18°C) in every month. The year-round warm temperatures occur because the sun's rays are never too far from being directly overhead in the afternoon, even in winter, because the latitudes are always close enough to the sun's direct rays.

The climate type that is found over the southern tip of Florida is a tropical wet-dry climate (map 2.1). The adjective "wet-dry" indicates that precipitation is abundant in some months but lacking in others. The wet months occur when the direct rays of the sun are closest (i.e., summer) because the heat from the sun causes warm, moist air to rise. This allows vertical cloud growth and on many days causes rain to fall from those clouds. In fact, cloudiness and precipitation prevent summer temperatures from being even higher than they otherwise would be. The dry months in southern Florida (October to April) occur when the overhead rays of the sun are farthest away and are focused on the Southern Hemisphere. During these months, south Florida is too far away from the direct heat of the sun for warm air to rise much. Instead, sinking air from a high-pressure system in the tropical Atlantic Ocean suppresses rising motion. At the same time, precipitation produced by mid-latitude frontal activity (which will be explained in more detail in chapter 4) usually does not extend far enough south to impact southern Florida significantly. This tropical wet-dry climate type is found in the southern part of the peninsula, south of Alligator Alley (I-75), and is characterized by very seasonal rainfall but little difference in temperature throughout the year compared to most places farther north.

The tropical wet-dry climate is also sometimes known as the tropical savanna climate because a type of natural vegetation called savanna tends to overlap with regions in this climate type. A savanna is a tropical or subtropical grassland with occasional tree clumps in the landscape. The distribution of trees in this type of grassland is controlled by subtle differences of a few feet in elevation. In slightly higher topography, drainage conditions are better and trees are able to thrive. Conversely, in lower elevations, where the soil is waterlogged, most trees cannot survive. The trees generally do not provide a closed canopy in a savanna environment, except perhaps in isolated areas such as a riverbank, where they can acquire water even during the dry season. Savannas evoke images of African safaris, but the southern peninsula of Florida can also be described as having savanna vegetation in some areas and grass-filled wetlands in other areas. In tropical wet-dry climates of southern Florida, Everglades wetland grasses grow quickly in the wet season and become dormant during the dry season. In the wet season, they recover quickly and resume growth. During the dry season, alligators in the Everglades lose their hiding places, so the hunting season generally ends in November in order to protect them and keep their population sustainable.

Subtropical Florida

The second generalized climatic type in Florida—subtropical—is found over most of the peninsula and across northern Florida (map 2.1), where the passage of cold fronts in winter causes average temperatures in January to dip below 64°F (18°C). The temperature regime in northern Florida varies too much from summer to winter to be considered a tropical climate, so most of Florida lies in the subtropics. Specifically, it can be said that the rest of Florida has a humid subtropical type of subtropical climate.

Unlike southern peninsular Florida, the humid subtropical climate part of Florida has a less distinguishable dry season. In summer, the abundant surface heating causes the warm air to rise and creates clouds and precipitation, just as in south Florida. These showers contribute heavily to Florida's abundant precipitation. In addition, the occasional tropical disturbance, tropical storm, or hurricane can bring rainfall to any part of Florida during summer or fall. However, unlike tropical wet-dry Florida, northern and central Florida are too far north to be affected as much by the sinking air from the high-pressure system that suppresses precipitation in the winter in south Florida. Winters tend to be rainier in northern Florida than in the tropical zone because as cold air moves southward toward the lower peninsula, the rising air motion along the leading edge, which is caused by the colder air over taking the warmer air and lifting it, decreases as the colder air loses its intensity and drier air infiltrates, which decreases rainfall amounts.

The boundary between the humid subtropical climate and the tropical wet-dry climate is not a discrete line, as depicted in map 2.1. Instead, there is a gradual transition zone between the two climatic regimes. About 60 percent of the annual average precipitation occurs from May to October (the rainy season) in northeastern coastal Florida at Jacksonville. In Melbourne, along the central Florida Atlantic coast, about 70 percent of the annual averaged rainfall occurs during these months, and in Miami, 75 percent of annual rainfall occurs during the rainy season from May to October.

Measuring Temperature

Temperature ranges are an important part of a climatology. Although electronic thermometers have now become the norm, the standard mercury-in-glass thermometers have been used for centuries to measure temperature. Special types of mercury or alcohol-in-glass thermometers are available that leave indications within the thermometer at the highest or lowest point the temperature has reached, similar to a bathtub ring. This special type of thermometer can be whirled around to remove the "ring." These instruments have long made it possible to know the daily maximum and minimum temperature; the observer simply resets the thermometer at the same time each day to know the maximum and minimum temperature over the previous 24 hours.

Traditionally, meteorologists calculated the average temperature for the day as the simple average of the maximum and minimum temperature for that day. For example, if the high was 90°F and the low was 70°F, the day's average temperature is 80°F. Of course, this isn't a very precise way to know the true average temperature during each minute of the previous 24 hours. In reality, on most days, the temperature is below the published daily average temperature for a longer time than it is above it, because the hottest part of the day is relatively short. Even though today we have computerized equipment that can measure temperature at many weather stations at every minute of every day, meteorologists continue to consider the daily average temperature in this crude manner—the average of the daily maximum and the daily minimum—so they can have an equal basis of comparison between today's temperatures and the temperatures of times before computerized equipment was available. Some temperature-recording stations still use either the mercury-in-glass thermometers or simple electronic devices that record only the maximum and minimum temperatures instead of more sophisticated computerized equipment that records more detailed records.

The average temperature for a month is computed as the average of the daily average values for each day in that month. The average temperature for the year is the average of the 12 average monthly values. The long-term average temperature for each of the 12 months of the year is based on many years of daily (and monthly) data. We can also compute the average annual temperature based on the average of each annual value. Typically, climatologists use 30 years of data to compute the climatic averages of an area.

Temperature Climatology of Florida

The average annual temperature for Florida increases steadily from north to south, from less than 66°F in the Panhandle to over 77°F near Miami (map 2.2). Considering that this difference of 11°F occurs along a north-south distance of approximately 360 miles, this gradient is about what we would expect, as the more northerly locations get less direct rays of sunshine than the peninsula and have winter temperatures that are less moderated by the waters of the Atlantic Ocean and the Gulf of Mexico.

Average Monthly Temperatures

Maps 2.3a and b provide a look at the average monthly temperatures across the state. During the winter months (December, January, and February), average temperatures range from those that require sweaters to those that allow short sleeves, dipping near or below 50°F in the Panhandle to above 68°F around Miami. The spring warmup in Florida occurs early in the year, just in time for students on spring break. During the spring season, average temperatures exceed 60°F almost statewide by March and statewide by April, and May values exceed 70°F statewide. The range in temperature is narrower in summer months (June, July, and August) across the state, as cold air rarely intrudes into northern Florida and cloud cover during the south's rainy season tends to suppress high temperatures and keep average temperatures in the low 80s

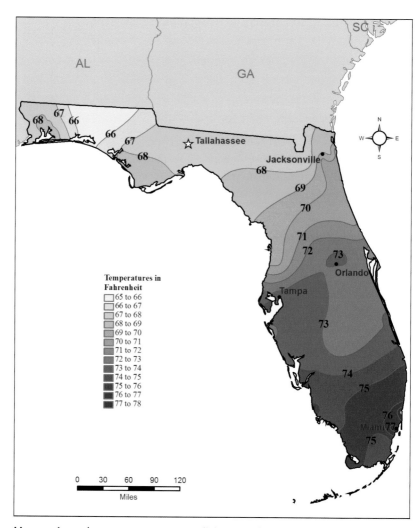

Map 2.2. Annual average temperatures (°F) across Florida, 1981–2010.

statewide. In September, some cooler temperatures return and averages in the 70s reappear. In October, the average temperature is in the high 60s in northern Florida. By November, a few northern stations have average temperatures as low as the high 50s. In December, temperatures in the 50s cover northern Florida, while in southern Florida, where temperatures are in the 60s and 70s, shirt-sleeves are still possible through the end of the year.

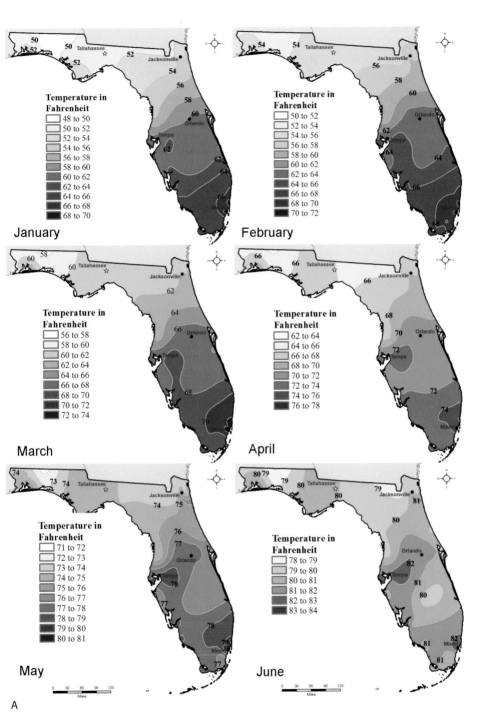

Map 2.3. *A*, Average monthly temperatures (°F) across Florida, January–June, 1981–2010.

(continued)

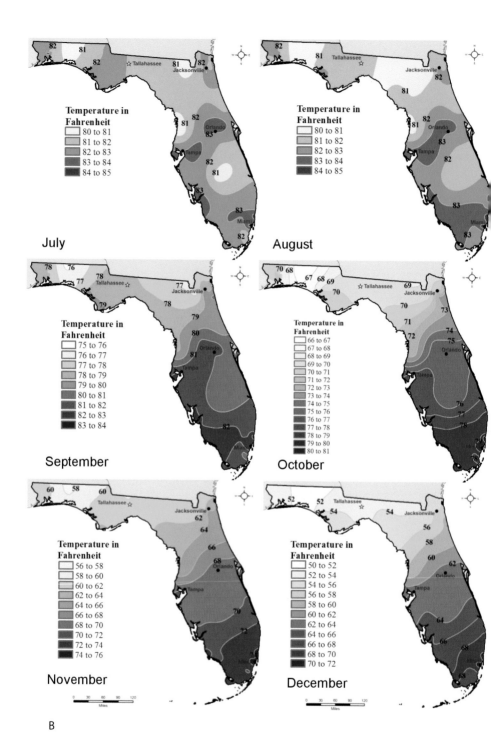

B

Map 2.3 cont. *B*, Average monthly temperatures (°F) across Florida, July–December, 1981–2010.

Average Maximum Temperatures

Map 2.4 shows the average maximum temperatures for January, April, July, and October, representing each of the four seasons. The average daily high for the whole year ranges from less than 79°F to over 84°F from north to south. January highs range from 60°F to almost 76°F, and April highs increase to 76–85°F. Maximum temperatures in July range only from 90 to 93°F and display an interesting pattern: higher temperatures in northern Florida. This is because sea breezes and cloud cover suppress afternoon high temperatures in tropical wet-dry Florida during the summer rainy season. The highest average maximum temperatures in July occur inland in central Florida, located away from the cooling effect of sea breezes that don't arrive until later in the day. Notice from the map how coastal locations have lower temperatures than inland areas during warm months such as July. October maximum temperatures range from below 80°F in the Panhandle to above 85°F in the lower part of the peninsula. By October, much of the tropical moisture has been pushed south of Florida as the first cool and drier air of the season enters Florida. Temperatures remain warmer over the peninsula though, from the influence of warm gulf and ocean temperatures, fewer clouds, and more direct sun.

Average Minimum Temperatures

The average minimum temperatures for January, April, July, and October are shown in map 2.5. For the year, the average daily minimum temperature is lowest in the Panhandle, where the average daily low is below 54°F. This hilly area is susceptible to the influences of cold air drainage as colder (and therefore denser) air flows into lower-lying areas. The average daily minimum temperatures increase slightly eastward and significantly southeastward from the Panhandle.

Even the relatively gentle topography of Florida's most rugged areas can produce noticeable influences on climate. Clouds act as a blanket that absorbs longwave radiation (heat) emitted upward from earth's surface and then re-emits some of it back downward, most noticeably at night, and lakes provide a warming influence on cool nights. The areas of cooler temperatures along the Georgia border and south of Orlando

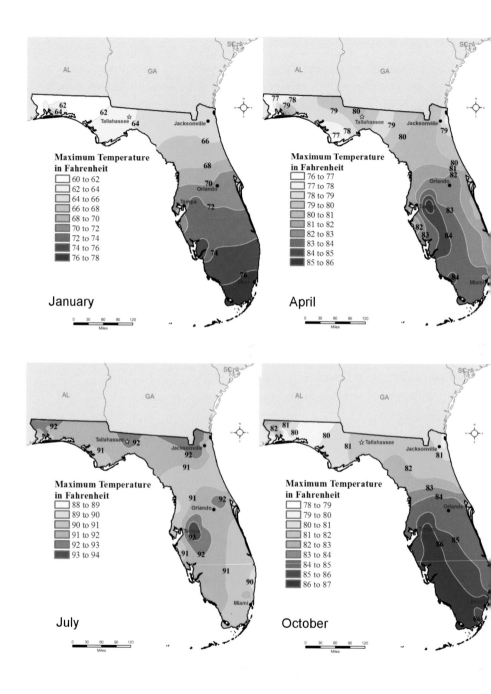

Map 2.4. Average daily maximum temperatures (°F) across Florida for January, April, July, and October, 1981–2010.

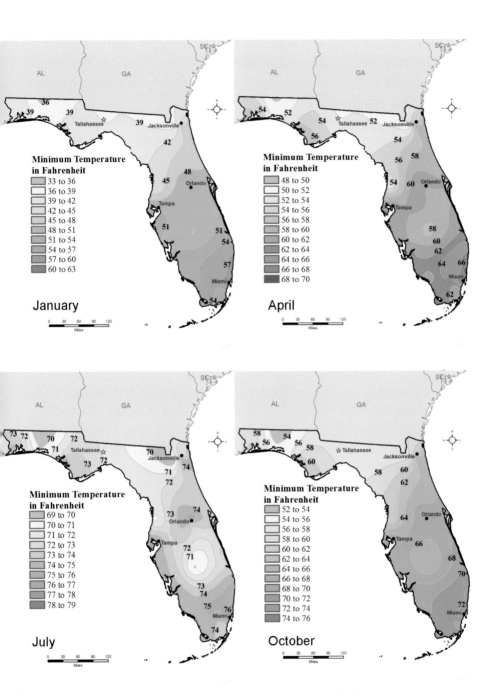

Map 2.5. Average daily minimum temperatures (°F) across Florida for January, April, July, and October, 1981–2010.

are prevalent because of terrain and less urbanization and because sinking motion in the atmosphere produces the clear inland skies that allow the earth to cool by emitting heat unimpeded by clouds and out to space through longwave radiation.

In January, average daily minimum temperatures are lowest (below 36°F) in parts of the Panhandle. The highest average low temperature in January is above 57°F near Miami, creating a statewide range of minimum January temperatures that exceeds 20°F. This range is among the largest in any state east of the Mississippi River. By April, the average daily minimum temperatures have increased statewide from the low 50s in the north to the mid-60s in the south, and the general increasing pattern from northwest to southeast prevails. Average minimum temperatures in July are quite high (low to mid-70s) and are similar across the entire state. The lowest temperatures are found inland, away from water molecules that create nighttime cloudiness that might absorb longwave radiation from the surface and re-radiate it back down onto the state.

Both April and October are transition months. During a typical April, temperatures are mild, but Floridians usually also get a preview of the upcoming summer heat for a few days. October is also a month of change; this is the month when the first cold front of the season typically pushes southward across the peninsula and brings in noticeably cooler and drier air. The pattern for October looks similar to the pattern for April but with slightly higher temperatures statewide. The summer heat tends to linger into October, as time is required for the cooler air masses to change the water temperatures that influence the air and surface temperatures. The inertia provided by relatively cool conditions of winter means that April is cooler than its fall analog month of October.

Florida Temperature Extremes

A look at the record highest and lowest temperatures ever recorded in Florida supports the idea that the northern part of the state is both hotter in summer and colder in winter than the peninsula. The highest temperature ever recorded in Florida was 109°F on June 9, 1931, in Monticello (Jefferson County), just east of Tallahassee (map 2.1). Many people would be surprised to know that because of the moderating effect of Florida's abundant nearby water bodies and humid air, this temperature record is the lowest all-time high temperature of any state except the six New England states, New York, Hawaii, and (not surprisingly) Alaska.

Florida's all-time lowest temperature in the meteorological record was a frigid -2°F in Tallahassee on February 13, 1899. Notice the proximity to the location of the all-time highest temperature! A similar pattern exists in the coastal state of Louisiana, where the locations of the all-time highest temperature and all-time lowest temperature are separated by only 40 miles, and, as is the case in Florida, both are in the farthest inland part of the state. Those bitter temperatures in Florida and Louisiana occurred during an Arctic outbreak that impacted the entire eastern United States from February 10–13, 1899. Not only was the all-time low temperature record for Florida set during this outbreak, but the low temperature records set during the same outbreak still stand in Louisiana, Ohio, and Washington, D.C. However, Florida's all-time low temperature record is warmer than all-time low temperatures for every state except Hawaii.

Fighting Freezes

In the past, citrus and winter vegetable growers in Florida used smudge pots to protect their frost-sensitive crops on cold, clear nights. In addition to warming the air directly, burning oil in smudge pots created a smoke layer that insulated the area near the surface of the earth. When the smudge pots were burning, less longwave energy escaped from near the surface and the crop could stay warm enough to survive. Nowadays, farmers spray water onto the crops to protect them from the cold temperatures. As the water freezes, it releases latent heat. This keeps plants just above freezing and warm enough to survive. Some people think that the coldest nights happen during a full moon, but this misconception is likely because they notice the moon more easily because of the lack of cloud cover on these cold nights.

Latitude and Sunshine

A few simple principles explain the overall patterns of temperature shown in the maps in this chapter. First and foremost, of course, is latitude. South Florida, which gets more direct sunshine, is warmer than northern Florida for most of the year. In the winter half of the year, the number of daylight hours is greater in southern Florida than in northern Florida. This gives the peninsula two latitude-related advantages: 1) more direct rays of sunshine; and 2) more hours in the day to receive these more direct rays. But the latitude effect also means that northern Florida has more daylight hours than southern Florida in the summer half of the year. This is one of several reasons why northern Florida has higher average temperatures than south Florida in the warmest months.

Cloud Cover

Latitude alone cannot explain all of the patterns in the temperature maps. A second important factor governing temperature is humidity and cloud cover. In the summer months when the sun is nearly overhead at noon, tropical wet-dry Florida is so cloudy that afternoon high temperatures are sufficiently suppressed to make southern Florida a bit cooler than northern Florida. In addition, coastal areas have cooling sea breezes and tend to have more humidity and cloud cover than inland Florida. Thus, the afternoon high temperature is typically cooler along the coasts. When Florida's maximum temperature record was broken on June 9, 1931, in Monticello, the skies were sunny and humidity was low at that inland location.

At night, the effect of humidity and cloud cover is the opposite of the effect during daytime hours. When all other factors are equal, a more humid and cloudier location will generally have warmer nighttime temperatures than drier places. This is because the shortwave energy that was stored at the earth's surface during the daytime hours can escape more easily to space in the form of longwave heat energy during the nighttime hours if there are fewer water vapor molecules in the atmosphere to absorb that energy and re-emit it back down toward the surface. Thus, humidity and cloud cover often act like a blanket at night, capturing more of the heat gained by the surface during the daytime near the surface. In Florida, as in many other places, the coldest temperatures generally occur on clear nights. It would be a safe bet that Tallahassee had clear skies on the frigid night of February 13, 1899.

Effect of Water Bodies on Air Temperature

A third factor that affects the temperature pattern is proximity to major water bodies. Large water bodies help keep the air warmer in the winter and cooler in the summer than in areas farther from the water bodies. Water and land have different heating and cooling properties. They retain heat differentially, and this affects the air temperature above them. In fact, a kilogram of land needs around five times less energy than a kilogram of water to increase its temperature by one degree Celsius. If the angle of the sun and cloud cover are the same over a land mass and a water body, the land mass will warm considerably quicker and reach a warmer temperature than the water body in the presence of sunlight. But water's ability to retain and slowly re-emit stored heat from summer and early fall in the form of longwave radiation means that in the winter and early spring, areas near larger water bodies will generally be warmer than areas located away from water bodies. These are important features to consider when studying Florida because it is surrounded by water on three sides. When a water reservoir is constructed, the adjacent neighborhood will generally be a little cooler in the summer and a little warmer in winter than it would be otherwise. The proximity of the stations that experience the high and low temperature records of Florida and the inland location of each should not be surprising considering the principle that inland locations tend to have more extreme summer and winter temperatures than coastal sites.

Urban Heat Islands

A fourth important principle regarding temperature is that large urban areas tend to be warmer than the surrounding areas. They are sometimes referred to as urban heat islands. This temperature pattern occurs for a number of reasons (figure 2.1). A summer walk barefoot across a blacktop surface serves as a quick and painful reminder that nonnatural surfaces such as asphalt, blacktop, stucco, and other building materials tend to absorb (shortwave) solar radiation more efficiently during the daytime hours than most natural surfaces. Also, many building materials absorb daytime heat and slowly release that heat throughout the night.

That heat then warms the surrounding urban atmosphere in the evening and nighttime hours. In addition, tall buildings can block breezes that would have helped to remove excess heat. Another reason for the urban heat island is that built-up areas have less vegetation than surrounding areas. Not only does vegetation provide shade, it also allows water to remain in the soil longer than would be the case if vegetation were not present. This happens because plants can regulate the uptake of water from the surface through their roots, stems, and leaf pores via transpiration, sending moisture into the atmosphere as water vapor. During wet times, plants can move more water from the soil to the atmosphere via transpiration, but during dry times, they can conserve water in the soil by reducing their rate of transpiration. Vegetation shades the soil and some of the sun's energy is used to evaporate moisture in the soil instead of being used to warm the surface. This prevents the blistering sun from raising temperatures quite as high as they would be on an unvegetated surface. When a vegetated area is cleared for development, much of the precipitation will run off the surface quickly instead of infiltrating the soil, where it could have later been slowly evaporated or transpired back up from the leaves of plants to the atmosphere. To minimize the flood hazard, rainwater in urbanized areas is deliberately channeled away from roads and buildings. Concrete is sloped into gutters on the sides of the road that lead to storm drains and culverts that prevent the water from seeping back into the urbanized surface. The water is then channeled to outlying areas using this method instead of using bioswales that temporarily retain the water as it filters into the ground. By moving the water out of the urban area quickly, the sun can then heat the urban surface more rapidly.

A final reason for the urban heat island is the waste heat produced by devices and equipment humans use. These include streetlights, fireplaces, air conditioner compressors, heat from industrial processes, and any type of motor. Because of the urban heat island, large cities can be more than 5°F warmer than the surrounding rural areas. The discrepancy is generally greatest in the evening hours. A few of the temperature maps presented in this chapter show some evidence of the urban heat island. For example, map 2.2 suggests warming near Miami, Orlando, Tampa, and Jacksonville.

Figure 2.1. Some causes of the urban heat island. Credit: Charles H. Paxton.

Florida's most prominent urban heat island is Miami, which is sand-wiched on a narrow strip of land between the Atlantic Ocean and the Everglades. This is visible in almost all of the temperature maps for this chapter. An island of warmer temperatures is usually evident around other major urbanized areas of Florida. For example, Jacksonville is located right at the 69°F annual temperature isotherm (line of equal temperature), while rural places just to the west at the same latitude are cooler than 68°F (map 2.2). After considering the importance of the urban heat island as a temperature booster, you may wonder why the coldest temperature Florida has ever recorded was in the city of Talla-hassee. But in 1899, Tallahassee's population was little more than 3,000. It is likely that it was the only weather station for many miles around, so it was crowned as Florida's all-time cold spot.

Think Local

A fifth influence on temperature distribution is simply local conditions. These local conditions can take many forms, including but not limited to the following examples. Your home will be cooler if it is shaded by a forest than it would be if it were next to an open field. A town may have slightly cooler temperatures if it is at a higher elevation than the next town down the road, even in the low hills of northern Florida. But at night, cooler air in those hills will drain downhill into the adjacent valleys because colder air is denser than warmer air. This is why on the coldest nights of the year, the citrus trees in the lower elevations are the most likely to suffer damage. The rooms on the side of your home that receive the afternoon sunshine will be warmer than those on the side of your home that receive the morning sun. A formerly swampy area that is drained in order to develop a subdivision will be likely to experi-ence a rise in climatological temperature, even if few trees are cut in the process, because a greater percentage of the sun's energy will be used to heat the surface instead of being used to evaporate water, as was the case when the area was swampy. These are but a few examples of how local conditions affect temperature.

A final factor that affects the temperature distribution in Florida, and elsewhere, is local and regional air circulation. When the wind at a location is blowing in from a warmer area, the temperature will be a little warmer than would otherwise be expected. Likewise, when the wind is blowing in from a colder area, the temperature will be colder than would otherwise be expected. In meteorology, winds are always named by the direction from which they blow. This is why a north wind in the winter (which is sometimes called a northerly wind when it is more variable in direction) generally brings cold air, at least in the Northern Hemisphere. Some places are climatologically more prone to have northerly winds, while others are climatologically more prone to have southerly winds. The same is true of ocean currents: some coastal places are more prone to having currents that flow north to south; these currents keep places in the Northern Hemisphere colder than they otherwise would be. Conversely, other places in the Northern Hemisphere are located next to currents that flow south to north; these keep those places warmer than they might otherwise be.

3

Tropical Breezes

The previous chapters described the incoming solar radiation, the seasons, and how temperatures vary. Solar radiation, which varies considerably from the equator to the poles, produces differential heating that ultimately provides the energy that drives the winds. The winds, in turn, move the air masses that create weather and climate across the globe and across Florida.

Florida's location, about one-third of the way to the North Pole from the equator, makes it subject to wind systems that are characteristic of the subtropical part of the earth. Some of the wind systems in the subtropics originate from the tropics and others are from the latitudes beyond the tropics. This chapter describes the origin of global wind systems and how they affect Florida. It does not cover the hurricane, the most famous type of organized tropical weather system that is steered by these global wind systems. That topic is covered in chapter 7.

Rising and Sinking Motion in the Atmosphere

The subtropical circulations that affect Florida are components of the global wind systems and are related to the belt of heated air that rises in the equatorial part of the earth. Near the equator, where the rays of the sun strike most directly, the surface of the earth is heated effectively. The atmosphere doesn't absorb much incoming (shortwave) solar energy, so much of the incoming solar energy heats the earth's surface, which then heats the air near the surface. This heated air rises because hot air expands and becomes less dense than the cooler air above. As the hot air rises and expands, it cools at a rate of 5.4°F per 1,000 feet with negligible net exchange of heat with the surrounding air. This expansion and contraction with no heat exchange is known as an adiabatic process. Adiabatic temperature changes are those in which cooling occurs solely because of the expansion of rising air and warming occurs solely because of compression of sinking air. Eventually the air will cool to the temperature at which the water vapor in that air cannot remain in vapor form. Then the water vapor in the rising air begins to condense and cloud formation begins. This is similar to what happens when you exhale hot, moist breath that cools and condenses as a cloud in front of your face on a cold day. Thus, paradoxically, the belt of latitudes around the world where the sun is shining most directly is often cloudy, especially in the afternoons, when air near the earth's surface warms, rises, cools, and condenses into clouds. When the water vapor condenses, the saturated adiabatic cooling rate is much slower than the unsaturated rate. Cloudiness acts as a moderator of the temperature because cloud cover can reflect additional incoming shortwave radiation from the sun. This rising motion near the equator triggers a major component of the atmospheric circulation around the globe.

Effects of Pressure

Winds generally blow from areas of higher pressure to areas of lower pressure. This is why it is important for meteorologists to measure and map pressure, the force of the air pushing down from above a point. Pressure always decreases with height in the atmosphere. If we drive up a mountain, less air is above us, the pressure decreases, and our ears will adjust to the change in pressure in the form of a pop. As air rises into lower pressure, the volume increases, expanding as it ascends, and the temperature cools. The opposite occurs when air heats. As air sinks into naturally higher pressures in the atmosphere, the compression of air caused by increasing weight (pressure) from air above it heats the sinking air. Within an area of high pressure where the air is sinking, adiabatic heating may cause a warm layer in the atmosphere that can be even warmer than the air below it that was not compressed as much; this is known as a subsidence inversion. Whenever temperature increases with height, such as in a subsidence inversion, air does not rise easily, so cloud formation is suppressed and conditions are likely to be fair, as we often see in high-pressure areas.

The Coriolis Effect

Whenever an enclosed area of high pressure exists, the natural tendency is for air to move radially away from the high-pressure center toward low pressure. However, once the air begins moving away from the high-pressure center, the rotation of the earth causes an apparent deflection of the air, to the right of the flow in the Northern Hemisphere and to the left of the flow in the Southern Hemisphere. This is known as the Coriolis effect; it is associated with the Coriolis force (map 3.1). The result is a clockwise and outward flow of air around Northern Hemisphere

Map 3.1. Effects of the Coriolis force for parcels moving north and south in the Northern and Southern Hemispheres.

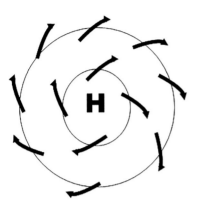 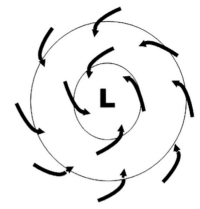

Figure 3.1. Surface wind flow associated with low- and high-pressure systems.

anticyclones (high pressure) and a counterclockwise and inward flow around Northern Hemisphere cyclones (low pressure) (figure 3.1).

Global Wind Patterns

The subtropical circulations that affect Florida are components of the global wind systems and are related to the belt of heated air that rises in the equatorial part of the earth. As the heated air on the surface of the earth rises upward to as high as 60,000 feet in the tropics, other surface air must move in laterally to take its place. Because of the effect of the earth's rotation on its axis, this movement of surface air toward the equator in the Northern Hemisphere comes from the northeast of the rising air. In the Southern Hemisphere, the rotation of the earth causes the air to come from the southeast. These winds blowing toward the equator are known as trade winds (figure 3.2), so named because they occur so consistently in the tropics that trade was conducted based on their presence. The records of the voyages of Christopher Columbus suggest that consistent winds from the northeast carried him from southern Spain toward the Caribbean Sea and northern South America. In Hawaii, the northeast trade winds are so consistent that businesses that face the northeast often have large open facades to take advantage of the natural breezy conditions. Many locations in the Southern Hemisphere have

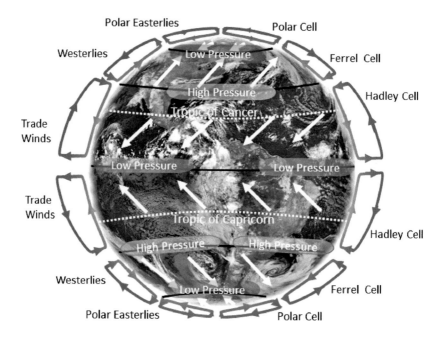

Figure 3.2. Global wind patterns.

southeast trade winds blowing from southeast to northwest that are just as consistent.

As the earth, which is tilted at 23.5 degrees on its rotating axis, orbits around the sun during the year, the belt of cloudiness that forms near the direct rays of the sun migrates seasonally within the tropics. Due to the earth's axial tilt, the most directly heated latitudes range between 23.5°N on June 20 and 23.5°S on December 21. However, the migration of this band of clouds generally doesn't fully extend to 23.5°N and 23.5°S latitude because by the time the surface gets heated each season, the sun's direct rays have already migrated a bit away from that latitude. The belt tends to extend farther away from the equator over large land masses—which heat up quickly and easily—than over water bodies.

When the direct rays of the sun are over the equator—near March 21 and September 22—the zone of rising air and cloud cover is near the equator and winds from the northeast collide into winds from the southeast, as shown in figure 3.2. When the sun's direct rays shine north of the equator—between March 21 and September 22—the zone of rising air is in the Northern Hemisphere and therefore is closest to Florida. During that time, southeasterly trade winds often cross the equator from the Southern Hemisphere, then veer to become southwest winds due to the rotation of the earth and become the counter-trade winds. The convergence of wind and moisture between the two prevailing circulating trade wind patterns has nowhere to go except up. The rising motion caused by air being heated is reinforced by this convergence. This belt of cloudy conditions is known as the intertropical convergence zone (ITCZ); it is shown in figure 3.3. It is intertropical because it oscillates northward and southward throughout the year but remains within the tropics, and it

Figure 3.3. Satellite image showing the ITCZ over the Pacific Ocean and (less distinct) over the Atlantic Ocean. Source: NOAA Geostationary Satellite Server (2015).

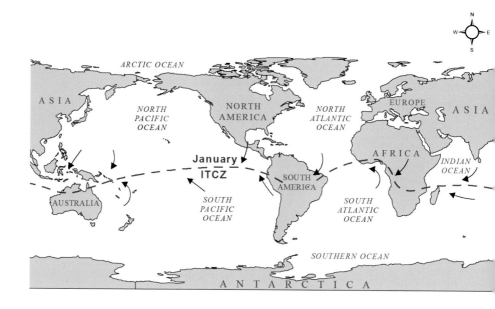

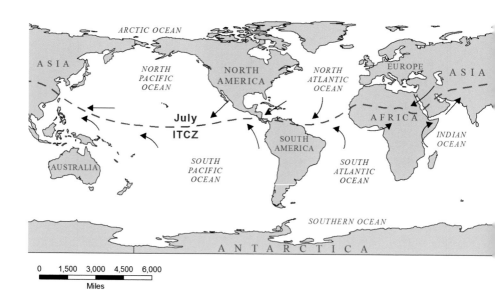

Map 3.2. Annual migration of the ITCZ.

involves converging trade winds along a zone around the equator (map 3.2). Pressure features such as the ITCZ and semi-permanent subtropical anticyclones are important for weather and climate not only because they enhance or suppress cloud cover but also because of the circulation associated with them.

The Hadley Cells

The hottest parts of the earth are generally not that close to the equator. Instead, they are at latitudes that are not far from the overhead rays of the sun but remain just out of reach of the cloud cover year-round. Additionally, the warm air that rises near the equator is part of a large circulation called a Hadley cell (figure 3.2), in which rising air drifts toward a pole and sinks, warming adiabatically, around 30°N and 30°S latitude. This sinking motion also causes the air to become drier from adiabatic heating. This is where fair weather associated with subtropical high-pressure areas over the oceans and hot places over the land such as the desert Southwest and the Sahara Desert tends to occur. Since Florida lies at similar latitudes to these places, why is it neither as hot nor as dry as these places? Part of the answer has to do with Florida's proximity to large water bodies, which moderate temperatures, but part has to do with other factors such as tropical circulations that bring cloud cover and local circulations of sea breezes.

The air at the ITCZ stops rising when it reaches the height where its buoyancy (which is caused because it has been heated at the surface) is no longer enough to allow it to continue rising. That is the height that some of its air begins to flow northward (in the Northern Hemisphere) and southward (in the Southern Hemisphere). This often happens at the height where temperatures increase with height in a warm layer known as an inversion. The tropopause is a particular inversion at the interface between the troposphere—the lowest layer of the atmosphere—and the stratosphere—the next higher layer. The tropopause can be close to 60,000 feet near the equator but decreases in height to as low as 22,000 feet near the poles. Pictures from space show how shallow the life-giving atmosphere is. In the middle latitudes, half the mass of the atmosphere is below 18,000 feet, or about 3.4 miles high.

In the Northern Hemisphere, at some height above the northeast trade winds, the atmospheric circulation is generally from the opposite direction—from the west or southwest. In the Southern Hemisphere, at some height above the southeast trade winds, the atmospheric circulation is generally from the opposite direction—from the west or northwest. These characteristic winds aloft and the surface trade winds beneath them, comprise the Hadley cells. There are two Hadley cells, one mostly or entirely in the Northern Hemisphere and the other mostly or entirely in the Southern Hemisphere.

The air on the poleward side of a Hadley cell sinks for sound physical reasons. After spending some more time flowing aloft, the air flowing from the west or southwest in the Northern Hemisphere (or from the west or northwest in the Southern Hemisphere) eventually becomes quite cold, and this, along with some contribution from the rotation of the earth, gives the air a tendency to sink (figure 3.2). This sinking air tends to occur about one-third of the way between the equator and the North and South Poles, near 30°N and 30°S. In June and July, when the ITCZ is situated north of the equator, the latitude of the sinking air is positioned northward in both hemispheres around 35°N and 25°S. The opposite is true in December and January, when the sinking air is displaced southward to about 25°N and 35°S.

Sinking air is associated with high surface pressure because the sinking exerts additional air pressure on the surface. The sinking or subsiding air creates adiabatic heating as the air compresses, and the warming tends to dry out the air. Subsidence minimizes the development of cloud cover, so sinking air is linked to fair weather in most cases. The sinking is more persistent in the subtropical parts of the earth over the cooler oceans away from the heated land, creating a semi-permanent zone of high pressure (called an anticyclone) over the subtropical oceans. The high-pressure areas are much weaker in the winter as areas of low pressure with associated cold fronts pass over the oceans. The oceans release their stored heat from the summer and support relatively warm air above them, which then rises and leads to the areas of low pressure. The subtropical anticyclones, in the Northern Hemisphere, strengthen and slowly migrate northward as June and July approach (map 3.3a)

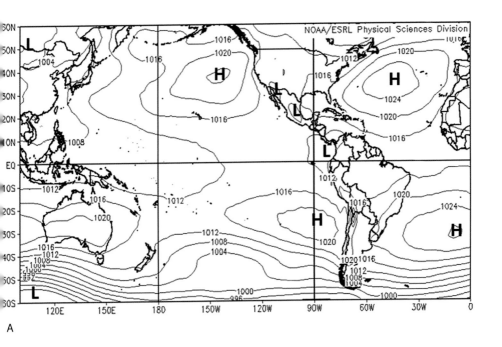

A

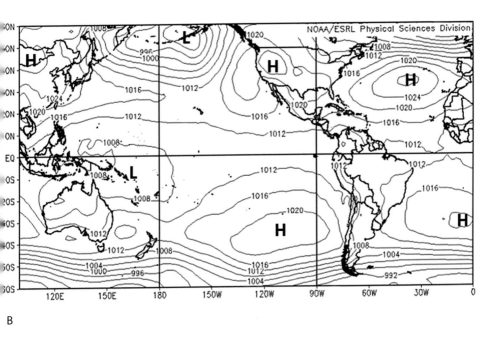

B

Map 3.3. Patterns of mean surface pressure in July (*A*) and January (*B*), 1980–2010. Source: NOAA Earth Systems Research Laboratory, Physical Sciences Division.

and southward as December and January approach (map 3.3b). Thus, semi-permanent high pressure dominates the climate of the subtropical oceans, especially in the summer. The subtropical anticyclones are so dominant that they persist through great heights in the atmosphere, even though they are most dominant over the ocean surface in the summer. Because of Florida's location, the Atlantic Ocean's subtropical anticyclone, the Bermuda-Azores high, contributes significantly to Florida's weather and climate and to the circulation of adjacent ocean waters.

The Polar and Ferrel Cells

Near the North Pole and the South Pole, the bitter cold air is so dense that it sinks, generating intense high pressure on the surface of the earth. That air then has no place to go except toward the equator. The cold polar air then moves southward (in the Northern Hemisphere) or northward (in the Southern Hemisphere) along the surface of the earth back toward the middle latitudes but is deflected by the rotation of earth so that it becomes an easterly (east-to-west) flow in both the Northern and Southern Hemispheres. This flow is known as the polar easterlies. Similar to what happens in the Hadley cell, as the air becomes relatively warm, around 60°N and 60°S, it rises and then moves toward a pole, cools, and begins to sink near the pole. The term Polar cell refers to this circulation cell, which is characterized by sinking air near a pole, easterlies, rising motion near 60°N, then flow from west to east but also back toward the North or the South Pole.

In the middle latitudes of both the Northern Hemisphere and Southern Hemisphere, the Ferrel cell is sandwiched between the Hadley and Polar cells. This is where the belt of prevailing west-to-east winds, called the westerlies, occur both at the surface and aloft beneath the tropopause. The middle latitudes within the Ferrel cell constitute a zone where weather systems intrude from both the poles and the tropics. In the upper part of the troposphere, just below the tropopause, between Polar and Ferrel cells, the polar jet stream tends to keep a balance in the atmosphere. Farther south, between the Hadley and Ferrel cells, the subtropical jet stream occurs. Both the polar jet stream and the subtropical jet stream play an important role in transporting energy and moisture, including storm systems, around the mid-latitude part of the earth.

The polar jet stream is usually located north of Florida. However, it can migrate across the Florida Peninsula during an occasional winter cold spell. By contrast, the subtropical jet stream is more likely to be directly above Florida.

The Bermuda-Azores High

As we have already seen, Florida's coastal subtropical location in the Northern Hemisphere means that its weather and climate are affected strongly by the Atlantic Ocean's semi-permanent subtropical anticyclone. This pressure feature is called the Bermuda-Azores high because its effects are felt near Bermuda on the west side of the Atlantic and near the Azores Islands on the east side (map 3.4). Americans sometimes shorten the name to the "Bermuda high," while Europeans and Africans tend to prefer the name "Azores high." The clockwise and outward flow of air around the Bermuda-Azores high has special significance for Florida. It causes warm, moist Atlantic and Caribbean air to infiltrate north and eastward to the peninsula and warm, moist Gulf of Mexico air to move northward to the Panhandle, especially in the summer, when the slowly migrating Bermuda-Azores high makes its northernmost approach (map 3.4).

Florida Wind Patterns

Since Florida lies to the north of the ITCZ throughout the year, the northeast trade winds are the tropical circulation that affects Florida most directly. Although Florida is a bit too far away from the tropics to experience the full effect of these winds, tropical trade wind flow is still important, particularly when the ITCZ makes its closest approach to Florida in the summer.

The winds favor certain directions in different areas of Florida. Wind roses are a way of showing the frequency of winds that are blowing from particular compass directions. Figure 3.4 shows annual wind roses for locations around the state. The length of each segment is related to the wind direction frequency, as indicated by the concentric ring labels. Moving from north to south, we see that each year Pensacola favors northerly winds, Jacksonville has a more balanced wind from all directions (hence

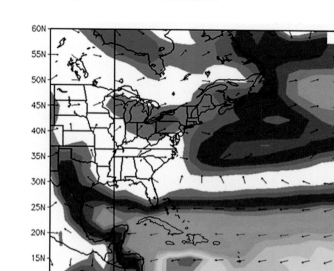

Map 3.4. Flow and speed of surface wind in the Bermuda-Azores high-pressure area, 1980–2010. Source: NOAA Earth Systems Research Laboratory, Physical Sciences Division.

the lower ring frequency), and Orlando winds come most frequently from the eastern half of the compass. As we head down the peninsula, we see an increasing trend of east winds from Tampa and Fort Myers; these winds become southeast in Miami. From one end of Florida to another, the winds change with the seasons. Map 3.5 shows the long-term average wind speeds and directions across Florida for each month. In January, the winds over the Panhandle are primarily from the north and then become northwest over northeast Florida. However, these winds are light; this indicates that they vary more in direction. Farther south, the January average winds become easterly and stronger. In April, the winds are more southerly over much of the state, which is not surprising as temperatures begin to warm up. In mid-summer, a clockwise wind flows from the high-pressure ridge that is typically stationed over central

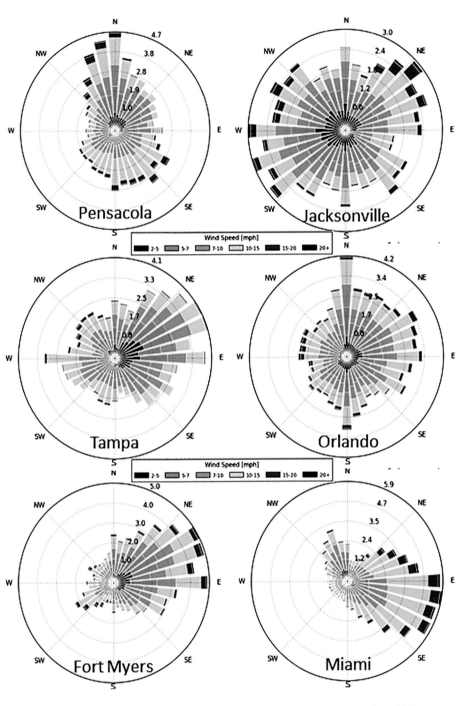

Figure 3.4. Annual wind roses for locations around the state. Source: Iowa State University of Science and Technology 2017.

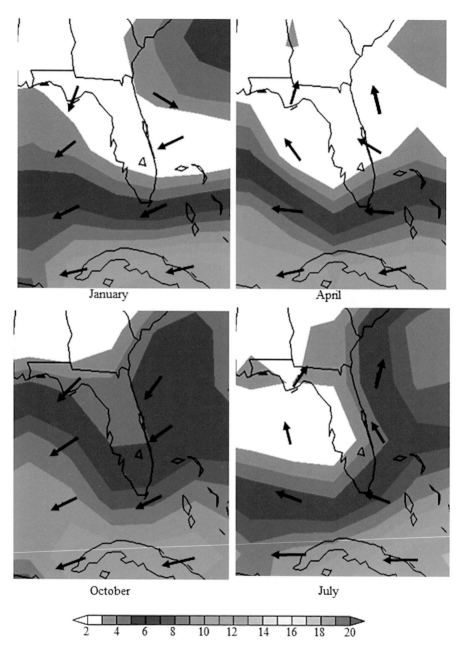

Map 3.5. Long-term average wind speeds (mph) and directions across Florida for January, April, July, and October, 1980–2010. Source: NOAA Earth Systems Research Laboratory, Physical Sciences Division.

Florida; southeast winds flow over south Florida and southwest winds flow over north Florida. As the seasons change again, October winds bring cooler air from the northeast across the state.

The Circulation of Land and Sea Breezes

The seasonal north-south oscillation of the ITCZ, trade winds, and wind circulation around semi-permanent pressure features such as the Bermuda-Azores high are considered global-scale wind systems because they affect a significant part of the world at the same time. However, many smaller-scale systems are also important in dictating the wind circulations that affect Florida. These smaller-scale circulations occur at the same time as the global-scale systems. Sometimes they are obliterated by global-scale systems and other times they amplify their effects.

The most prominent local- to regional-scale circulations for Florida are the sea-breeze and land-breeze circulations. These phenomena are important in most coastal areas of the world. On a warm afternoon, the land area along a coast warms much more effectively than the adjacent coastal waters. The warm air over the land rises, while the relatively cool air just offshore sinks. The sinking air offshore is linked to a small-scale high-pressure system that is much smaller in extent than the Bermuda-Azores high. The rising air over the land is linked to small-scale low pressure over Florida (figure 3.5). The near-surface air moves from higher to lower pressure, creating winds that blow inland from the Gulf of Mexico and Atlantic Ocean. Since winds are always named by the direction from which they blow, this phenomenon is known as a sea breeze (even though a gulf, ocean, or other body of water produces the effect).

At night, the land cools off much more rapidly and extensively than the adjacent water just offshore. The relatively cool air sinks, leaving local- to regional-scale high pressure over the land and corresponding lower pressure over the water, which stays almost as warm as it was in the daytime due to the ocean's ability to retain the heat it gained during the daytime hours. When this happens, the sea breeze circulation no longer exists; instead, the air flows outward from the higher pressure over land toward the lower pressure over the water. Because winds are

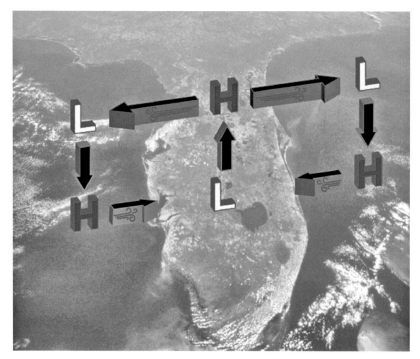

Figure 3.5. Sea-breeze circulation. Credit: Anita Marshall.

always named by the direction from which they blow, this circulation near the surface from land to sea is called a land breeze (figure 3.6).

Notice that in the course of 24 hours on a typical day, the circulation associated with a daytime sea breeze eventually weakens and then stops sometime in the nighttime hours, when the air over the land and the air over the water are around the same temperature. Then the circulation reverses itself to become a land breeze (in the night and early morning), when the air over the land becomes cooler than the air over the water. This circulation reaches maximum intensity when the land air is much cooler than the marine air. The circulation then weakens and stops when the air temperature over the land and the air temperature over the water once again become nearly equal (in the mid-morning). A sea breeze develops again as the land heats up (in the mid- to late morning) and reaches peak intensity during the late afternoon and into the evening. The intensity of a sea breeze or a land breeze is generally proportional to the degree of temperature difference between land and sea.

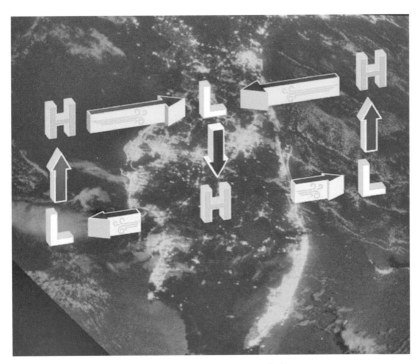

Figure 3.6. Land-breeze circulation. Credit: Anita Marshall.

Of course, the "sea" could actually be a lake if the lake is large enough, but in those cases, the phenomenon is known as a lake breeze. Sea breezes, land breezes, and lake breezes circulate at a small enough scale that the Coriolis force does not significantly deflect their wind direction. The terms sea-breeze front and lake-breeze front do not refer to an actual weather front like a cold front; they refer to a much smaller-scale feature. These terms refer to the leading edge of the sea (or land) breeze circulation, which typically only penetrates a few dozen miles in from the shore. However, at times such circulations can move all the way across the Florida Peninsula.

Sea and land breezes have been studied for centuries, dating back at least to the ancient Greeks, because they are well known to have strong impacts on everyday life. For instance, if a motorized boat is unavailable, a fisherman will wake up early enough to ride the land breeze out to sea. Then, after several hours of fishing, the sea breeze kicks in to bring the fisherman back to shore. Air quality is also affected by the sea and land breeze circulations. The sea breeze will blow daytime pollutants from a coastal city inland in the daytime hours, but when the land breeze is in force, it will return the pollutants to the coastal urban area and then offshore.

Air temperatures in coastal zones are also affected by these circulations. In some places, nocturnal temperatures sometimes drop much more in areas adjacent to the coastline than elsewhere because the persistent land breeze can actually move the top layer of surface water that has been warmed by daytime heating well offshore. Then, colder water moves upward through the upwelling of deeper water to the near-surface layer. The result is a nighttime low temperature that in some cases can be even colder at the coastline than farther inland, where the land surface cools the temperatures. In Chicago, which is affected by the land and lake breezes from Lake Michigan, nighttime lows often plunge low enough to overtake the urban heat island effect of the city because of this upwelling effect.

Florida is one of the world's most prominent locations for sea and land breezes. The resort town of Gulf Breeze near Pensacola was named for the sea breeze that is such a welcome relief from the sultry summer conditions. Peninsular Florida experiences the effects of the sea and land breeze circulations from both the west and east at the same time. A satellite image on a typical summer afternoon usually shows evidence of the double sea breeze in the form of a line of cloud cover just inland from the coast on both sides of the Florida Peninsula (see figure 3.7). Evidence of the sea breeze is also often visible just inland from the Panhandle. The leading edge of the sea breeze, the sea breeze front, moves moisture-laden air inland but slows because of increasing friction from the terrain, forests, and buildings. Air from the sea continues to move in faster behind the sea breeze front, creating horizontal convergence, or a pileup of air just inland from the coast. The piled-up, moisture-laden air is forced upward, where it condenses into a cloud that stretches parallel to the coastline.

Although sea breezes can occur at any time of year in Florida, the most noticeable months are May through September, when the most extreme heating over land occurs. Typically, a sea breeze occurs when the land is 5–10°F warmer than the ocean. The Florida sea breeze circulation

Figure 3.7. Satellite image of the double sea breeze. Source: NASA (n.d.).

can develop in both humid and dry conditions, but the greatest effect on Florida weather takes place in periods when strong pressure features are absent and the steering winds are weak and humid conditions persist. During such times, sea breezes can produce conditions that are favorable for producing or increasing vertical cloud growth into thunderstorms and more precipitation across coastal Florida. During the months of May through September, the surface heating of the Florida landmass is strongest and the sea breezes create rising motion that generates cumulonimbus (i.e., thunderstorm) clouds that bring lightning, hail, damaging winds, waterspouts, and tornadoes.

Another factor in the strength of the sea breeze is the depth of the atmosphere in which the vertical part of the circulation can take place. This is called the mixed layer (figure 3.8), which is characterized by vertical, turbulent motion. It is generated largely by the warm air near the surface that rises because it is less dense than surrounding air. The mixed layer is capped by air that sinks under the impacts of gravity. The height of this layer varies by time of day, time of the year, and other atmospheric conditions. If the depth of the mixed layer is low, then the sea

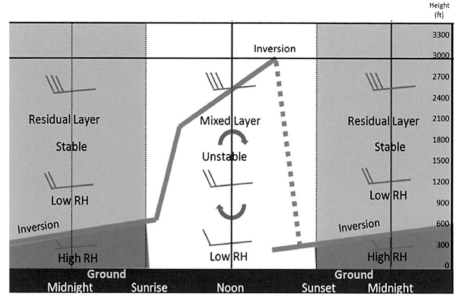

Figure 3.8. Evolution of the mixed layer. The solid line that becomes a dashed line represents the top of the mixed layer. The darker shades indicate night. Shaded areas indicate higher relative humidity (RH). The wind speeds are shown as barbs. Each barb represents 10 mph; half-barbs represent 5 mph.

breeze circulation will not progress inland very far. The mixed layer is shallowest in the morning when the winds are light, as the wind barbs in figure 3.8 indicate. Early morning is also when the relative humidity is highest. The mixed layer exhibits rapid rise as daytime heating takes its effect and is highest in the afternoon. The higher the mixed layer, the deeper the sea breeze circulation, which may be 5,000–6,000 feet (i.e., still well within the lower part of the troposphere) in the warm season. Above the mixed layer is the residual layer, where the mixing of air near the surface is negligible. As the sea breeze moves inland, the air rises along the boundary, then moves laterally back toward the water and sinks over the cooler water surface (see figure 3.5).

The fully developed sea-breeze circulation pattern has surface areas of high pressure over both the Gulf of Mexico and Atlantic Ocean adjacent to Florida, low surface pressure along the spine of the peninsula, and a divergent interior high-pressure area above the land. At the surface, air flows from the sea to the land, and aloft air flows in the opposite direction from the high pressure above the land toward low pressure over the Gulf and Atlantic Ocean, thus establishing the west coast and east coast sea breezes. It is important to note that the "high" pressure aloft is not higher than the pressure near the surface; it is high only in comparison to the pressure at the same level adjacent to it.

Sea-Breeze Modifications

As in other places that are affected by sea and land breezes, several factors can weaken or eliminate the circulation of sea breezes in Florida. Most prominently, the presence of a larger-scale circulation such as a tropical storm or a cold front passing through the area will wipe out the tendency for the sea breeze circulation to exist. In addition, dry upper-level air will tend to hinder the development of the coastal cloud cover linked to sea breezes, as moisture that would have formed the clouds along the sea breeze front is evaporated into the dry air as the cloud begins to form. Similarly, a decreasing depth of the mixed layer will suppress the circulation of sea and land breezes.

Other conditions can complicate the development regions where sea and land breezes circulate. For instance, the shape of the coast can alter the "typical" circulation pattern of sea and land breezes, particularly under changing wind regimes. When the winds are light, sea and land breezes tend to develop perpendicular to the coast and the associated clouds tend to be aligned parallel to the coast. This allows for the development of convex, concave, and straight-line sea-breeze fronts (see figure 3.9). Convex coastlines, like the coastline near Bradenton, can create a focusing effect that creates more wind convergence and stronger thunderstorms. We would expect extra precipitation in this area, and indeed, the area inland from Bradenton, at Myakka River State Park, receives more precipitation and lightning in the summer thunderstorm season than locations closer to the coast. The presence of estuaries and bays, such as those near Tampa Bay and Charlotte Harbor, also complicate the pattern of circulation of sea and land breezes.

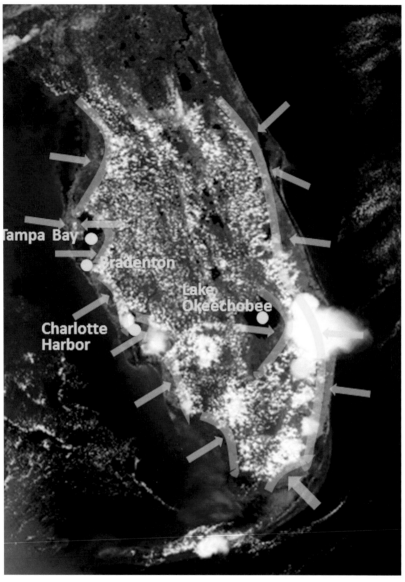

Figure 3.9. Effects of Florida's coastline on sea breezes. Source: adapted from NASA Worldview.

Lake Breezes

Lake Okeechobee in the southern portion of the state is large enough to create a breeze of its own. This Okeechobee breeze sometimes interacts with the sea breezes of the east and west coasts (figure 3.9). In the area west of Lake Okeechobee, a gulf breeze blowing from the west often converges with a lake breeze blowing from the east in the late afternoon. At the same time, east of Okeechobee, an ocean breeze blowing from the east converges with a lake breeze blowing from the west. The result is greater summer rainfall amounts east and west of the lake. Thus, in this part of south Florida, there is actually a quadruple circulation of sea breezes! Of course, the nocturnal land breeze can take on similar quadruple form, with winds blowing in opposite directions to create land-breeze fronts.

Wind Patterns that Affect Sea Breezes over Florida

As we have seen, the large-scale wind flow affects the wind patterns of sea breezes, and these patterns change the patterns of convective clouds over the Florida Peninsula. These wind patterns influence the direction of movement of sea breezes on Florida's east and west coasts and, in turn, affect precipitation patterns, particularly in the summer thunderstorm season.

One common pattern is characterized by an easterly (east-to-west) wind flow that causes the east coast sea breeze to move faster and farther inland than the west coast sea breeze. On these days, there is an early development of convection along the east coast. A merger of the east coast sea breeze and the west coast sea breeze takes place in the middle or west of the center of the peninsula, where most of the strongest convection occurs.

A less common wind pattern occurs when a stronger easterly wind flow is present. Convection begins along the east coast and the strong easterly wind flow quickly progresses toward the west coast. The east coast sea breeze and west coast sea breeze converge over the western portion of the peninsula and create strong convection and precipitation along this zone of convergence. In this regime, the strong convection happens later in the day (but usually still before 6 p.m.) and the most rain falls in the southwestern portion of the peninsula. The rain is focused on this area partly because of the concentration and enhancement of convection due to the convex curvature of the coastline in this area. In this pattern, precipitation is usually strong but brief.

Another common wind pattern occurs when a dominant westerly flow causes the west coast sea breeze to move farther inland than the east coast sea breeze. Convection starts much earlier on both coastlines than is the case in either of the previous two regimes mentioned. In this steering wind pattern, the west coast sea breeze propagates inland while the east coast sea breeze, which is pushing into the westerly flow, cannot penetrate more than a few miles inland. As a result, by 3 p.m., much of the peninsula is convectively active and growing clouds become thunderstorms. From 3–6 p.m., the east coast sea breeze remains near the east coast, while the west coast sea breeze has progressed to the center to the peninsula. From 6 to 9 p.m., most of the thunderstorms are along the east coast and the interior of the state. With this wind pattern, clouds cover a more extensive area of the state and thunderstorms do not dissipate until a later time of day than is the case with either of the wind patterns previously mentioned.

A fourth and final type of wind pattern is the "bucket" that holds all the other wind patterns that can have an impact on convective clouds over the Florida Peninsula. This wind pattern is composed of many different types of sea breeze days; these are influenced by tropical systems, nearby cold fronts, mid-latitude "short waves" that can generate feisty storms, and other disturbances. As a result, sea breezes and precipitation on these days can be highly variable.

El Niño and La Niña

Subtle changes in the trade winds governed by the subtropical areas of high pressure over the oceans have global impacts. At times, the trade winds are stronger than normal. Other times, they are weaker than normal. These patterns occur on an irregular two- to three-year cycle. Weaker easterly trade winds over the Pacific Ocean near Peru in South America will fail to blow the top layer of equatorial water westward, creating a pileup of warm water near the coast of Peru. This pattern is

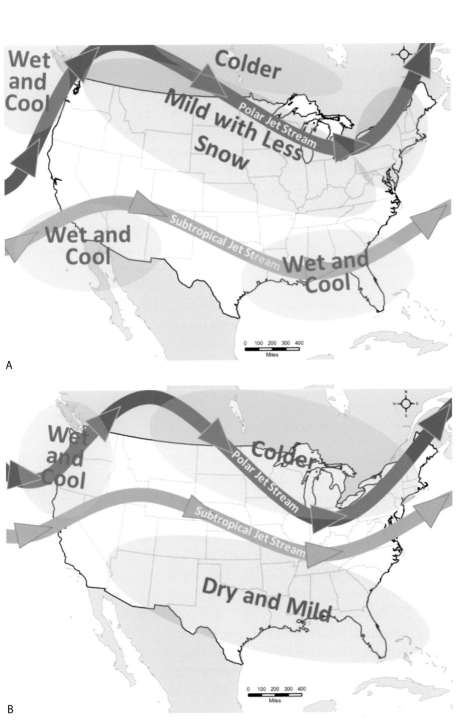

Map 3.6. *A*, Typical El Niño conditions; *B*, Typical La Niña conditions.

known as El Niño. When this happens, cooler-than-normal water exists toward Australia and Southeast Asia (map 3.6a). Conversely, when the trade winds are stronger than normal across the Pacific Ocean, the equatorial waters off the coast of Peru are cooled as warm surface water is pushed westward, while on the other side of the Pacific, the waters warm (map 3.6b). This pattern is known as La Niña.

El Niño and La Niña have global weather impacts and can change Florida's weather patterns. El Niño conditions tend to cause the jet streams of the Northern Hemisphere to flow farther south, which increases wind shear (change of wind speed or direction with height). This helps amplify thunderstorms along cold fronts and creates more favorable conditions for strong tornadoes over Florida in the winter. For tropical weather systems, El Niño has the opposite effect. Tropical cyclones, including hurricanes, tropical storms, and tropical depressions, are sensitive to wind shear and usually weaken in the presence of strong shear. During La Niña, because the jet stream has retreated northward, Florida winters are drier and the hurricane seasons tend to be more active.

Human Influences on Wind Flows

Anthropogenic (human-induced) changes in land cover can also impact the land and sea breeze systems. In 1999, a group of meteorologists used computer modeling techniques to compare the weather in Florida in the years 1900, 1973, and 1993 (Pielke et al. 1999). In the year 1900, Florida had not been developed extensively and the Everglades were largely untouched. The computer models used land-use maps from 1900 while simulating the development and movement of sea and land breezes. The team then compared the results of that model to the actual observed weather conditions in 1973 and 1993. They found that the differences in land cover resulted in a reduction of deep cumulus rainfall that caused a 9 percent decrease in precipitation in 1973 and an 11 percent decrease in 1993. This would tend to support the notion that anthropogenic changes in land use are playing a role in Florida's weather and climate.

4

Fronts and Winter Weather

Television meteorologists describe how cold fronts, warm fronts, and sometimes occluded fronts will change the weather. A front is the zone of temperature change between different air masses, typically between warm and cold air. Sometimes the difference in temperature is extreme as the front passes. Cold fronts are the leading edge of a moving mass of cold air and warm fronts are the leading edge of a mass of moving warm air. Fronts are associated with areas of low pressure that swirl in and up. Cold air masses are usually associated with air around a high-pressure area that is sinking and spreading out.

Meteorologists recognized the concept of a front around the time of World War I. This segment of clashing air masses reminded them of a zone where two armies meet, so they borrowed the term "front" from the military type of front on a battlefield. The weather associated with cold fronts, warm fronts, and stationary fronts tends to be fairly distinctive. In this chapter, we will explore the processes that take place along these weather systems and the mechanisms by which they produce significant weather in Florida.

Cold Fronts

Outside the tropical parts of the earth, precipitation in the form of rain, snow, and thunderstorms usually occurs along a linear or curvilinear frontal zone where cold air meets with much warmer air. If the cold air is advancing and thus is pushing the warm air away, generally toward the equator, it is called a cold front. When a cold front passes over an area, the temperature often falls suddenly. When cold air masses move over oceans or large bodies of water, the water releases its stored heat to the overlying air, causing the cold air to become warmer and gather moisture. The coldest and driest air to impact Florida behind cold fronts travels from the polar regions toward the equator over land areas covered by snow, where the air mass undergoes little modification.

Air masses with different temperatures have different densities; colder air is denser than warm air, and as it moves, will force the less dense, warmer air up over it (map 4.1). The lifting or rising motion created by the steep wedge of cold dense air along a cold front can spur development of strong thunderstorms.

In the summer, the temperatures warm over the polar areas and because there is less contrast in temperature from north to south, the cold fronts are weaker. Cold fronts rarely intrude deep into Florida during the summer because the Bermuda-Azores high ensures that tropical air remains dominant and the more direct summer sun shrinks the polar air masses far to the north of Florida. Summer cold fronts cause temperatures to drop in places such as Chicago and New York, but generally not in Florida. But as winter approaches, the sun's rays are lower in the sky, the number of daylight hours is shorter, and the northern air masses get colder. These temperature differences within the atmosphere create stronger steering winds, which push the cold air toward the equator and toward Florida in the broad mid-latitude zone where bitter cold air meets much warmer and moister air. This zone becomes the battleground where strong and persistent thunderstorms frequently develop. After a cold front passes through Florida, high pressure behind (i.e., to the north and west of) the front moves in. High-pressure areas usually bring clearing skies and have clockwise winds that flow downward and out from the center.

For Floridians and people in most other areas of the country, the cold air usually comes from the interior of the continent, so after a cold front passes through a location, the temperature drops and the incoming cold air is drier than the tropical air it replaced. The moisture that is lifted into the atmosphere at a cold front generally produces a short period of heavy

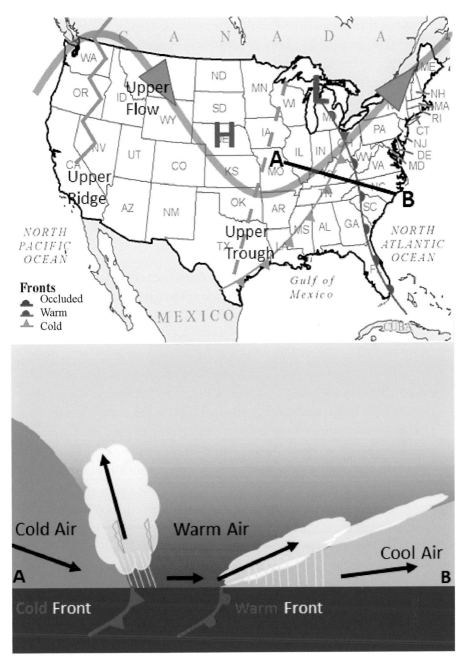

Map 4.1. Cold, warm, and occluded fronts and upper-level flow depicted on a weather map and cross-sections of cold and warm fronts.

rain and then the air dries out, the skies clear, and temperatures fall. The clear skies and low humidity in high-pressure areas after a strong cold front passes through often bring the coldest nights in Florida.

Once or twice a year, Floridians experience what is called a back-door front. These occur when the steering flow pushes a cooler air mass toward the southwest instead of to the southeast or east, as usually occurs. In Florida, back-door fronts bring a cool, moist, air mass that has been modified over the waters of the Atlantic Ocean. Back-door fronts can sometimes be driven in a westerly or southwesterly direction by the flow of a northeast trade wind that is farther north than usual.

Warm Fronts

As high pressure passes and low pressure approaches, the winds become southerly and warm air moves northward. The lighter, less dense warm air is wedged up over the cold, dense air mass that is retreating toward the pole. This is known as a warm front. Unlike cold fronts, warm fronts have a broad and gradual uplift zone. The warm air tends to move more laterally as it rises (map 4.1). This results in an extensive layered cloud cover that is more blanket-like and widespread in coverage than what we see with cold fronts. As a warm front approaches an area, the broad lifting motion creates large areas of cloudiness and persistent rainfall from nimbostratus clouds. These clouds occur in large sheets and are easily recognizable by their dull, gray, flat appearance. If precipitation is occurring, it tends to be of light to moderate intensity and cover a large area. Intense rain, lightning, and other forms of severe weather are less common along warm fronts than they are along cold fronts. As the warm front passes, temperatures begin to increase and the sky may begin to clear. Because warm air that affects Florida must come from either the Caribbean, the Gulf of Mexico, or the Atlantic Ocean, the warm air masses are also humid. The passage of a warm front over a location usually means that temperatures and tropical moisture are increasing. Anyone who has lived in or visited Florida is all too familiar with how these warm air masses feel.

Stationary Fronts

Sometimes steering wind patterns near fronts become weak, and for a time neither the cold air mass nor the warm air mass are moving and displacing each other. When this happens, the front becomes stationary. Stationary fronts are often marked by a band of clouds along the front. These fronts tend to have similar characteristics to whatever type of front stopped moving. Thus, if a cold front slows down and barely advances for several hours, all locations under it may be experiencing weather that is characteristic of the weather a cold front produces. If a warm front slows down and barely advances, the places around it may be experiencing weather similar to what a warm front produces, at least until the front resumes its motion. Locations that are experiencing rain are likely to continue to do so until the stationary front resumes its motion. Thus, persistent stationary fronts are commonly associated with flooding.

Occluded Fronts

Occluded fronts form when a cold front overtakes a warm front and another distinct boundary is formed between the coldest air behind the cold front and the cool air mass to the north of the warm front (map 4.1). When this happens, the warm air that is moving poleward behind the warm front is occluded or lifted above the ground and away from the low-pressure center at the earth's surface.

The weather associated with occluded fronts often includes precipitation. But because the occlusion doesn't occur until well after the cold and warm fronts have begun supporting precipitation, the most intense precipitation and violent weather has usually already occurred before the occluded front forms. Occluded fronts are usually associated with old storm systems that have already expended most of their latent energy.

Weather Maps

Meteorologists examine many different types of weather maps that depict conditions at different levels of the atmosphere. The most typical weather maps show patterns near the surface of the earth and include an analysis of the various types of fronts based on plotted data. Cold, warm, occluded, and stationary fronts are usually easy to identify by symbols and their abrupt temperature and wind changes. Fronts usually trail from a traveling area of low pressure (a cyclone) that moves generally from west to east across the middle latitudes. Map 4.1 shows a weather map in which both warm and cold fronts are moving across the United States. The cold front is drawn as a blue line with blue triangles pointing in the direction the cold air is moving, which is usually eastward and toward the equator. Warm fronts are drawn using a red line with red semicircles pointing in the direction the warm air is moving, which is generally eastward and poleward. Occluded fronts are depicted as lines with alternating purple triangles and semicircles on the same side of the purple line. Stationary fronts are shown as lines that alternate red semicircles on one side of the line pointing poleward and blue triangles on the other side of the line pointing equatorward. On the lines that depict stationary fronts, the blue triangles point in the direction the cold air is trying to push; red semicircles point in the direction the warm air is trying to push. But neither air mass is moving much. Fronts may drape across hundreds of miles and may be cold at the easternmost end, stationary in the middle, and warm on the west end.

Low-pressure troughs with no discernable temperature discontinuity are depicted on weather maps as dashes that are sometimes colored brown. These troughs are not fronts but are areas where the wind may change direction and the weather may be active. Sometimes air on one side of the trough is much moister than the air on the other side. This type of trough is known as a dryline.

Analysis of Surface Weather Maps

Before weather balloons and radar were first used (in the 1940s) and before satellites were first launched (in 1960), meteorologists could make weather measurements only near the ground. Weather observers would look up at the sky and estimate the amount of sky cover, cloud height, visibility, and type of precipitation. The observers would then use weather instruments to measure the temperature, atmospheric moisture, wind direction and speed, precipitation amounts, and pressure. Early in

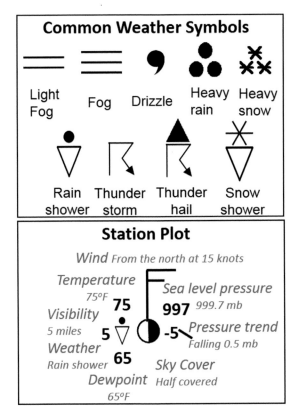

Figure 4.1. Common weather symbols and a weather observation station plot.

the twentieth century, meteorologists created a standardized code and a technique for plotting surface observations on a weather map. Figure 4.1 shows a weather observation station plot as it would look on a surface weather map and some common weather symbols. Because atmospheric pressure varies with height, the pressure measurements were adjusted to what the measurement would be at sea level. Once a map was plotted, typically by junior meteorologists, it was handed over to senior meteorologists, who would analyze fronts and discontinuities denoted by sharp temperature gradients and wind shifts. Typically, the analysts would then interpolate the pressures between the plotted observations and draw lines of constant pressure, called isobars, that denote high- and low-pressure areas. The closer the isobars are, the greater the pressure gradient and the stronger the wind. Analysts would also draw lines of constant temperature, called isotherms, to highlight where cold and

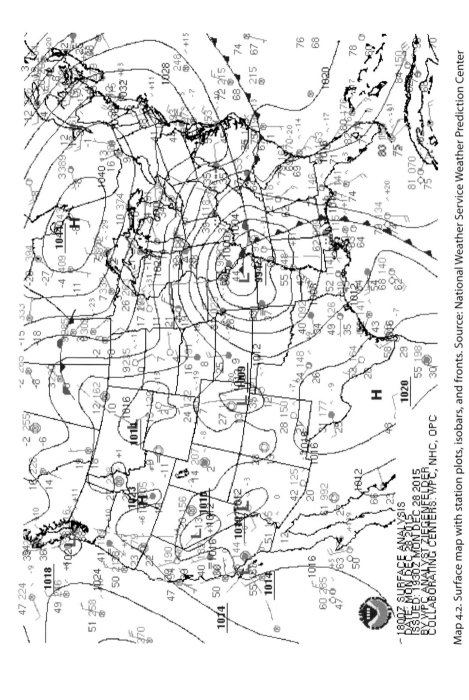

Map 4.2. Surface map with station plots, isobars, and fronts. Source: National Weather Service Weather Prediction Center

warm air masses were situated. Nowadays, computers plot the observations and perform much of the analysis. Map 4.2 shows a weather map that a computer plotted and analyzed.

Mid-Latitude Cyclones

Fronts are part of a larger system called a mid-latitude cyclone, an area of low pressure that extends upward into the atmosphere. In the Northern Hemisphere, the air in a mid-latitude cyclone spirals upward, inward, and counterclockwise. The cyclone develops as fronts form at the boundaries of air masses of different temperatures. Recall from map 4.1 that cold fronts and warm fronts tend to be draped from the enclosed area of low pressure (called a cyclone), so the fronts and the cyclone itself are all components of the mid-latitude cyclone system. From figure 4.2, it is evident that the cold and warm fronts take on the appearance of a wave, and since cold air and warm air tend to meet in the middle

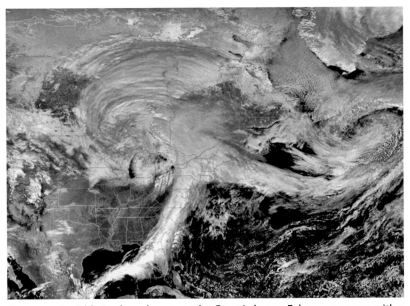

Figure 4.2. A mid-latitude cyclone over the Great Lakes on February 21, 2014, with a cold front extending southward across Florida and a warm front stretched toward the Atlantic Ocean that becomes a cold front wrapping into another mid-latitude cyclone. Source: NOAA Environmental Visualization Laboratory (2014).

latitudes, the name mid-latitude cyclone makes sense. Notice how the cyclones over the Great Lakes and the Atlantic Ocean in figure 4.2 are connected by a front.

Air circulates around cyclones counterclockwise in the Northern Hemisphere. Thus, to the west of a cyclone, cold polar air is pushed toward the southeast as a cold front. To the east of a cyclone, warm tropical air is pushed toward the northeast as a warm front. The entire system (cyclone and fronts) is pushed eastward by the prevailing mid-latitude westerly wind flow at heights between 10,000 and 40,000 feet. The counterclockwise flow around the low-pressure area, such as the one centered over the Great Lakes area in map 4.1, pushes the cold front toward the southeast and the warm front tends to move toward the north or the northeast in the mid-latitude cyclone. Florida is on the fringe of the middle latitudes and is in an area where these cyclone systems might develop and have a direct impact on the state. Usually, though, the centers of these mature traveling low-pressure systems are north of Florida.

The cold fronts that cross Florida typically originate hundreds or even thousands of miles northwest of the state. They usually take several days to sweep across the country before they finally arrive over Florida. In the summer, the cyclonic centers are so far north that even the fronts trailing from them cannot reach as far south as Florida. Florida is firmly entrenched in the warm air, usually far from the boundary between warm and cold air. But in the winter, as the cold pool of air extends farther toward the equator, it is quite common for Florida to be in a position where fronts affect the weather. Northern Florida is affected more directly and for a larger part of the year than southern Florida. This is because cold fronts originate farther north and often stall or weaken by the time they reach central and southern Florida, especially because the clockwise flow around the western side of the Bermuda-Azores high-pressure system allows the tropical air to maintain its position and resist movement toward the equator. The strength and position of the Bermuda-Azores high differs from week to week and from year to year, but it generally weakens and retracts south and east in the winter. Thus, cold fronts can advance deeper into the peninsula in the winter. Cold fronts are mostly a winter phenomenon for Floridians. But the term "cold" is relative; it only means that the air behind the polar front is much colder than the air ahead of it. Thus, in the spring or the fall, a cold front may bring nighttime low temperatures of 50°F or even 60°F to Florida, while a winter cold front can cause temperatures to dip below freezing.

For mid-latitude cyclones to develop, several things must happen. Air masses of contrasting temperature, and troughs of elongated low pressure in the middle and upper troposphere must be upstream above the air masses as low pressure develops at the surface and intensifies. As mid-latitude cyclones develop (in a process called cyclogenesis) and intensify (in a process called deepening), the area of low pressure extends up but tilts back toward the west in higher regions of the atmosphere. As distinct cold or warm air masses develop and move, the corresponding frontal development is called frontogenesis. As the cyclonic system moves along at a speed and direction dictated by the steering wind, many types of weather phenomena are possible, depending on temperatures and amount of atmospheric moisture. Typically, after the peak intensity of a mid-latitude cyclone, an occluded front will form as the cold front overtakes the warm front. When this happens, another distinct boundary is formed between the coldest air behind the cold front and the cool air to the north of the warm front as warm air that is moving toward the pole behind the warm front is occluded or lifted above the ground. The action of a mid-latitude cyclone is similar to the action of a blender, and eventually the low-pressure system is aligned vertically and the air masses mix so much that the cyclone weakens and dies.

The Steering Flow

In addition to supporting counterclockwise winds around it, the low-pressure area and the fronts trailing from it are being pushed at the broader scale. This circulation generally comes from the upper-level steering flow of the mid-latitude westerlies that were described in chapter 3. The entire cyclonic system—the low-pressure center, its counterclockwise circulation, the warm front, and the cold front—usually tracks from west to east or from southwest to northeast because it is steered by westerly (i.e., west-to-east) wind patterns that are typical of the middle latitudes. Although these winds, of which the polar jet stream is a part, tend to move from west to east, they can dip toward the south or the north along their general west-to-east trek. Thus, fronts generally move from west to east as they wrap counterclockwise around the low-pressure center. This is analogous to water in a stream, which can whirl around in eddies even as the eddies themselves move in the more general path from the source of the stream to its mouth.

Weather Balloons and the Depth of the Atmosphere

As the earth moves around the sun and seasons change and the earth rotates on its axis creating day and night, the temperatures are never balanced. The atmosphere is continually in motion, trying to seek equilibrium by mixing between high- and low-pressure areas. As we see, equilibrium is never accomplished.

Pressure is the downward force of the atmosphere or simply the weight of the atmosphere above. In physics the term "force" refers to the product of mass times the downward acceleration due to gravity. As we go higher into the atmosphere, the pressure decreases. Meteorologists must consider how the different levels of the atmosphere are changing and impacting the weather that we see and feel every day. The fundamental part of making a forecast is assessing the current weather, which involves understanding the present and future distribution of pressure, temperature, humidity, winds, and other variables. Measurements of these important variables are taken around the world in many locations near the ground.

Weather balloons are launched once or twice daily at official stations around the world (figure 4.3) and carry radiosonde instruments up to an altitude where pressure approaches zero (about 20 miles). These send data using a radio signal. The balloon ascends at a rate of about 1,000 feet per minute and because of decreasing pressure the balloon grows larger. Eventually the balloon pops and the radiosonde instrument makes a parachute landing. The observed vertical data are plotted on weather maps at standard pressure levels and analyzed to determine patterns of the relevant variables in the atmosphere. The data from the measurements are also assimilated into numerical model runs that simulate future behavior of the atmosphere. The weather maps from the surface to the upper levels change from day to day as weather systems develop, migrate, and expire.

Figure 4.3. A weather balloon with a radiosonde and a parachute. Credit: Charles H. Paxton.

Table 4.1. Height of pressure levels

Pressure level (in millibars)	Approximate height above sea level (in feet)
1,000	300
925	2,500
850	5,000
700	10,000
500	18,000
400	24,000
300	30,000
250	34,000
200	39,000
150	45,000
100	53,000

Weather maps of the surface typically depict pressure, wind, and moisture patterns. Pressure always decreases with increasing height, and maps of areas above the surface typically depict the height of standard pressure values, temperature, moisture, wind, and the amount of spin (vorticity) in the wind flow. These data are vital to our understanding and forecasting of weather and climate. Table 4.1 shows the standard (mapped) pressure levels in the atmosphere, measured in millibars (mb), with the corresponding approximate heights in feet.

The Skew-T Diagram

Radiosonde measurements of the atmosphere are typically plotted on a graph called a Skew-T diagram that is often referred to in National Weather Service Area Forecast Discussions that are available to the public (figure 4.4). The Skew-T illustrates the temperature, the dewpoint temperature (the temperature to which the air would need to cool for the humidity to reach 100 percent, a data point that indicates how moist the air is), and wind speed and direction that are plotted from the surface up to the 100 mb pressure level. These graphs are important for forecasting weather. Forecasters look for areas where the temperature and dewpoint temperature are close, indicating a cloud layer, or where the temperature

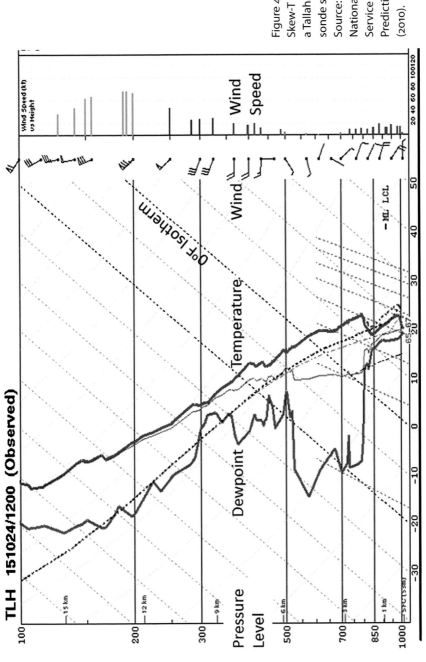

Figure 4.4. A Skew-T diagram of a Tallahassee radiosonde sounding. Source: NOAA and National Weather Service Storm Prediction Center (2010).

and dewpoint are far apart, indicating little humidity and no clouds. The graph has horizontal lines that represent pressure levels and slanted (skewed) lines that represent temperature. Dotted lines depict the mixing ratio, the mass of water molecules divided by the mass of all other molecules in the atmosphere.

Other lines on the graph help with the analysis of rising motion in the atmosphere. A dashed green line represents the dry adiabatic lapse rate, or the rate of cooling of unsaturated air. A curved green line shows the moist adiabatic lapse rate, which represents the rate at which saturated air is cooling. Meteorologists can follow these lines on the Skew-T diagram to determine whether it is likely that convective (vertically oriented) clouds will develop and how vigorous the convection might be. The temperature and dewpoint profiles on the Skew-T also provide indications of whether rain, snow, sleet, or freezing rain might occur. The winds are plotted along the right side to show how they change with height and at which level stronger winds associated with jet streams might be. Historically, aviators needed to know how high they needed to fly so they would know if they could utilize the jet streams or if they needed to fly at a lower level to avoid the jet streams.

The Middle and Upper Levels of the Atmosphere

Meteorologists usually look at flat weather maps but must imagine the weather occurring at a very shallow depth, only about 10 miles deep, on our rotating planet, which is a sphere. Using computers, meteorologists can derive many other atmospheric parameters to make a more thorough assessment of the atmosphere. On upper-level maps (called charts), such as the 500 mb chart in map 4.3, atmospheric pressure is the same across the entire chart and the mapped feature on constant pressure charts is height. The solid lines on map 4.3 represent lines of equal height. On upper-level charts, areas of high heights correspond to higher pressure and areas of lower heights correspond to lower pressure. Away from the frictional effects of earth's surface, the atmosphere generally flows parallel to the solid lines on upper-level charts, with higher heights to the right of the flow (in the Northern Hemisphere).

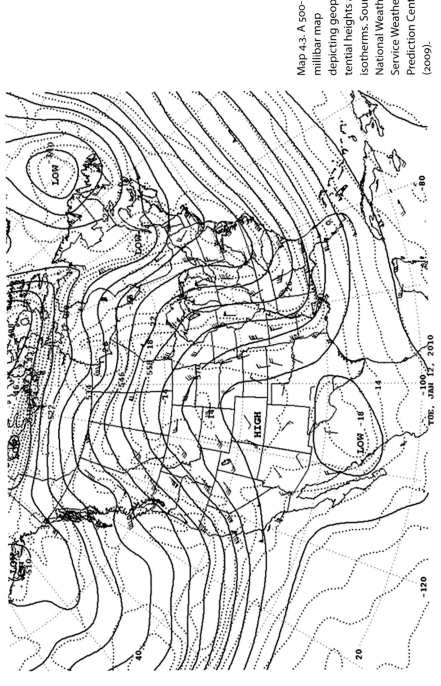

Map 4.3. A 500-millibar map depicting geopotential heights and isotherms. Source: National Weather Service Weather Prediction Center (2009).

On upper-level charts, elongated areas of high heights (i.e., high pressure), as is the case from Colorado north into Canada on map 4.3, are called ridges. Ridges are sometimes depicted by a zigzag line on a chart. Likewise, areas of extended low heights (i.e., low pressure), as is the case with the lines that dip toward the south along the Atlantic seaboard and over the Pacific Ocean in map 4.3, are troughs and are sometimes depicted by a dashed line on a chart. Because air in the Northern Hemisphere flows clockwise around high pressure and counterclockwise around low pressure, it is easy to see that the air in map 4.3 or in any other standard height map always flows clockwise around the ridge and continues flowing parallel to the solid lines until it becomes wrapped up in the counterclockwise trough to its east. Thus, the flow in upper-level maps is approximately west to east, but it jogs toward the pole around ridges and jogs toward the equator from the ridge axis to the axis of the next downstream trough.

The system of ridges and troughs is continuous around the Northern Hemisphere. The flow encircles the North Pole from west to east as it jogs north to south or south to north between a ridge and its adjacent trough. On any given day, there are usually three to six pairs of ridges and troughs around the Northern Hemisphere and another three to six pairs of ridges and troughs around the Southern Hemisphere. The amplitude and position of these ridges and troughs change from day to day. Map 4.4 shows a hemispherical view of the 500 mb height map. Typically, the amplification or de-amplification builds up gradually over the course of a few days, but this sometimes happens faster when storm systems develop. Thus, meteorologists can identify the trend in amplification over the course of a few days and make a reasonable guess about the type of flow to expect in the near future. The migration of the axes of ridges and troughs are a bit trickier to predict, but often the axes will propagate toward the east by a few hundred miles per day. Sometimes, though, a ridge or trough will move toward the north or the south (with or without amplification), and sometimes it may even move from east to west, in a process called retrogression. It is also common for a ridge-trough configuration to be locked in place for a week or more.

The position of the ridges and troughs at the 500 mb level and above is critical for understanding and predicting outbreaks of cold air. This

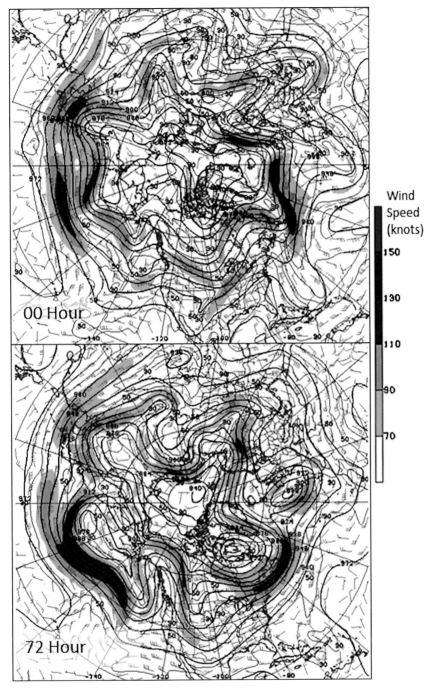

Map 4.4. Hemispherical views of the progression of the 500-millibar geopotential height map for 72 hours. Dark areas depict stronger wind. Source: National Weather Service NCEP Central Operations (2016).

is because flow on the ridge-to-trough side of the wave brings polar air toward the equator and is the key ingredient that causes polar outbreaks in the location just downstream from the ridge-to-trough side. Thus, cold air outbreaks in Florida will nearly always require a strongly amplified 500 mb ridge over western North America and an equally deep 500 mb trough over eastern North America, as is the case in the scenario depicted in map 4.3.

The Pacific-North America (PNA) Pattern

Atmospheric scientists have devised a clever way to describe the amplitude and position of the major ridge-trough configurations over North America. The term Pacific-North America pattern (PNA) refers to this ridge-trough pair. A PNA index of 0.0 represents a ridge and trough of normal amplitude and position; the normal condition is a weak ridge over the Rocky Mountains and a weak trough with an axis over the Ohio River Valley. A positive PNA index represents conditions in which the ridge and trough are both more amplified than usual. During times when the ridge-trough pair are weaker than normal or are even inverted so a trough exists over the Rocky Mountains and a ridge is present over eastern North America, the PNA index is negative. If the PNA index is as extreme as ±1.0, the pattern is one standard deviation above (or below) the average pattern, meaning that it is in the highest (or lowest) 16 percent of all the cases at the corresponding time of year that are in the period of record. If the PNA index is as extreme as ±2.0, then the pattern is two standard deviations above (or below) the average pattern, meaning that it is in the highest (or lowest) 2.5 percent of all the cases at the corresponding time of year that are in the period of record. The historical monthly values of the PNA index are readily available online from National Oceanic and Atmospheric Administration's Climate Prediction Center and on other websites. It can be inferred that Florida's historically cold winters must have occurred at times of a very high PNA index.

Jet Streams and the Flow of Polar Air

The two primary jet streams that encircle the Northern Hemisphere and have an impact on Florida are the polar and the subtropical. Both of these jet streams migrate and vary daily and seasonally. The polar jet is responsible for guiding cold air masses into Florida and is embedded within the ridges and troughs, including those associated with the PNA pattern. Its typical west-to-east orientation drifts toward the south in the winter. When this wavy pattern dips toward Florida, a colder period can be expected. If the dip is pronounced and if it is associated with a frigid air mass, cities as far south as Miami may have freezing conditions.

To the south, the subtropical jet may work in tandem with the polar jet to bring cold air masses into Florida, but is higher and does not undulate as much as the polar jet. Thus, the subtropical jet has less influence on winter cold snaps in Florida. But when the subtropical jet is near Florida, it has a major effect on amplifying the potential for severe thunderstorms during the cool season. A cold air mass can be rapidly displaced toward the south and can reach Florida over a period of several days. The faster the steering flow, the greater the likelihood that the cold air mass will retain its frigid characteristics by the time it reaches Florida. If a large section of the upstream United States is already experiencing a snow-covered winter, the air mass can retain its cold characteristics more effectively. The Panhandle receives the full impact of the displacement of a cold air mass before it moves south to the rest of the state and is modified by the Gulf of Mexico and the Atlantic Ocean.

Fronts and Severe Weather

In Florida, as is the case in most other locations, tornadoes, damaging winds, and hail are usually found along or near cold fronts where strong lifting motion occurs. That is because as the cold air displaces the less dense warm air, the warm air is lifted up into the atmosphere. This lift occurs relatively easily, since the warmer air is less dense, and the ease with which the warmer air rises allows clouds to grow vertically to great heights, especially when abundant moisture is present in the warmer air mass. Of course, abundant atmospheric moisture is often available over Florida. So along cold fronts, it is not uncommon for Floridians to see tall cumulonimbus (thunderstorm) clouds, heavy rain, lightning, and hail. The zone of lift that cold fronts create is usually narrow, so precipitation usually ends quickly, but in some cases a zone of lift may occur in a linear zone up to a hundred miles ahead or sometimes behind the frontal boundary. Vigorous fast-moving lines of storms that parallel the frontal boundary are known as squall-line thunderstorms and can be more severe than those that accompany the front itself. If the counterclockwise winds around the cyclone are particularly strong at upper levels of the atmosphere, the tall cumulonimbus clouds may begin to rotate, leading to stronger, faster, and more persistent thunderstorms. The availability of so much moisture around Florida typically means that the approach of a cold front in which significantly colder air is replacing much warmer air is a signal that thunderstorms are imminent.

Winter Weather

At first glance, it may seem odd that winter weather should be an important part of a discussion about Florida's weather. But often the most drastic weather impacts occur when places experience rare weather phenomena. Florida's winter weather can wreak havoc because farmers, operators of vehicles, and homeowners are all too often ill prepared to deal with harsh winter weather, such as freezing temperatures, snow, ice, sleet, and freezing rain. This section explores how each of those weather phenomena occur, their frequency and intensity in Florida, and some of their many impacts.

Cold fronts trailing from mid-latitude low-pressure cyclones begin sweeping through Florida around October, but these fronts seldom have frigid air behind them. These fronts do little more than bring a narrow line of showers or thunderstorms. Sometimes they bring a large area of rain and then drier, slightly cooler air during an otherwise relatively dry winter. As is the case with all cold fronts, the colder air wedges under the warmer and more humid air, lifting it to create thunderstorms or at least rain. Sometimes during the winter a shallow layer of cold air associated with high pressure will move over Florida behind a weak cold front in a regime known as overrunning. As the high pressure moves east, the wind begins to blow from the Atlantic Ocean, but upper winds from the opposite direction overrun the cold air and create a cold day with a thick blanket of clouds and persistent areas of rain.

Advection Freezes

Advection freezes occur when cold air crosses Florida and creates freezing conditions. Occasionally an Arctic air mass is steered toward the south, creating a particularly intense cold front that can bring devastating low temperatures to Florida in the winter. Bitter cold high-pressure air masses develop during the winter, when darkness reigns over the snow- and ice-covered Arctic areas from Siberia to Canada. These are driven by an upper-level steering flow characterized by an intense high-pressure ridge over the western United States and a similarly amplified low-pressure trough over the eastern United States. Map 4.3, which records data from January 12, 2010, depicts the height in the atmosphere where the pressure is 500 mb. As these steering currents amplify, they acquire the ability to transport the frigid air mass south into the United States. The dotted lines in the figure show temperature in degrees Celsius, and the wind barbs indicate the measured direction and speed at radiosonde sites. This transport of Arctic air into Florida is known as an advection freeze and may persist under strong northerly surface winds for several days. Large low-pressure systems associated with winter storms north of Florida may also rapidly push (or advect) cold air into the state. Windy conditions are common during advection freezes.

Radiation Freezes

When cool and dry high-pressure air settles over Florida behind a cold front, the ground will begin to lose heat as the sun sets. The heat radiates from the surface of the earth into space at night in the form of long-wave radiant energy. This is known as radiational cooling. If it is cold enough, it is known as a radiation freeze. Potential radiation freezes are affected by areas of moisture and clouds within the atmosphere. Clouds act as a blanket for radiational cooling by keeping warmth trapped near the earth's surface; the result is warmer nighttime temperatures and a lower chance of a radiation freeze. Strong winds in the lower layers of the atmosphere also reduce the chance of radiation freezes by creating mixing that impedes the cooling effect of longwave radiant energy from the surface back out to space.

Radiation freezes, which are most likely to occur under clear, calm night skies, can create frost that can kill tender vegetation. A killing frost can occur even when air temperatures are above freezing because some objects, such as roofs, windshields, and plant leaves, radiate heat efficiently enough to cool below freezing even if the air above it is warmer than freezing. The impact of radiational freezes on vegetation and crops can be mitigated more easily than the impact of windy advection freezes. This can be done by warming the air or by providing a constant spray of water to keep plant temperatures just above freezing. However, during advection freezes, attempts to warm the air are often ineffective because a fresh supply of frigid air is constantly being blown into the area.

Historical Freezes in Florida

In the infamous winter of 1898–1899, a series of cold fronts swept down from Canada through the Great Plains and into the Gulf region. Several successive low-pressure systems developed and moved along the frontal boundary and brought heavy rain to the Gulf region and heavy snowfall to areas north of the front. On February 13, 1899, a low-pressure system intensified off the southeastern coast of the United States. This system brought two inches of snow to Jacksonville and measurable snow as far south as Fort Myers. That evening, under the frigid air brought by the Arctic high pressure behind the cold front, Tallahassee reported a temperature of -2°F, the lowest temperature ever recorded in Florida.

Significant freezes impact Florida every 10–20 years and typically occur when the El Niño/La Niña cycle is neutral. However, freezes have occurred during El Niño and La Niña conditions when the atmosphere provided a clear path for Arctic air to aim for Florida. During the 1980s, three uncommon major freezes struck Florida and devastated the citrus industry. After this, many orange and other citrus grove owners moved their operations farther south. While most freeze events only last one day, that is long enough to inflict serious damage on winter vegetable crops, citrus groves, livestock, and even marine life. The loss of coral and other marine life to severe cold temperatures can have profound impacts on the ocean environments, and recovery from such an event can take years.

More recently, a severe cold outbreak in the winter of 2010 had damaging results across the country. Florida experienced record-breaking low temperatures that had not been seen for many decades. The most remarkable characteristic of this winter was not its record low temperatures but its unusual duration. Much of central and northern Florida experienced subfreezing temperatures for 12 days, and inshore water temperatures dropped below 50°F for 10 consecutive days. The winter (December, January, and February) average PNA index was an unprecedented 1.77, and the PNA for every month in that winter was also well above normal. The Miami office of the National Weather Service released a statement that said that "a blocking pattern in the middle and upper levels of the atmosphere allowed for a continued flow of Arctic air to sweep across the eastern half of the country" (NOAA, National Weather Service, and Weather Forecast Office, Miami 2010). The first cold wave did not arrive until early January, when two successive Arctic cold fronts moved into Florida. Multiple daily low temperature records were broken and/or tied during this event. Areas such as Gainesville and Apalachicola experienced air temperatures below 26°F. A record in Florida was set for the number of successive freezes in a single month. Tallahassee recorded 14 freezes in January alone. Frozen precipitation was seen across central Florida during the third week in January as a series of mid-latitude cyclones swept across the country amid the cold conditions. The storm systems were carried by an active subtropical jet stream that stretched across the country. On February 12, 2010, a news report released from New York was titled "Color America White." On that day, 49 of 50 states had snow on the ground; Hawaii was the only state that did not have snow.

The 2010 record cold event took a huge toll on crops and tropical plants in Florida, and over $500 million was lost in crop damage. Many coral reefs in Florida were impacted when they were exposed to water temperatures well below their reported thresholds of 61°F; the result was rapid coral declines of unprecedented severity. The frigid weather conditions that dropped water temperatures to critical thresholds below 50°F also caused unprecedented harm to sea turtle and manatee populations in the shallow waters surrounding Florida. Approximately 5,000 sea turtles and hundreds of manatees were impacted by the cold. They became lethargic and were stranded; many died.

Hazardous Winter Storms

Winter storms that drop freezing and frozen precipitation are an even more formidable hazard than bitter cold temperatures. Fortunately, Florida seldom experiences significant impacts from dangerous storms that produce these conditions, although the state certainly did in 2010. To understand how sleet, freezing rain, and snow form, it is important to recall that whenever a cold and warm air mass meet, the warmer air will be pushed up in the atmosphere as the colder (and therefore denser) air undercuts it.

Unlike most other severe weather associated with fronts, significant winter weather such as sleet, freezing rain, and often even snowfall tends to be more common in association with warm fronts than along cold fronts. Air that rises above the warm front, as is depicted in map 4.1, may cool to subfreezing temperatures on ascent as its moisture condenses to form clouds. Thus, any winter precipitation that falls from those clouds is likely to start as snow, even if the warm front is over Florida. But as the snow falls, it may descend through warmer air ahead of the warm front. If that warm air layer is thick, then the snow will melt and hit the ground as rain. However, if the snow falls into temperatures around 32°F or colder, it will remain as snow when it hits the surface. The latter is uncommon in Florida because the state seldom has a deep enough cold layer to support snow.

But there are two other possibilities. If the layer of air with temperatures above 32°F is thin and is sandwiched between two subfreezing layers of air, the snow can melt on descent through the warm layer but then re-freeze on descent through the cold layer near the surface of the earth. In such a case, the snowflake is transformed to a rain drop and then to a small pellet of ice much smaller than a hailstone. This type of precipitation is called sleet. It occurs next to a zone where snow is occurring.

Worse yet is freezing rain, which happens when rain falls into a shallow freezing layer of air near the surface and freezes on contact with the surface. Freezing rain is very dangerous for automobile drivers, who may lose traction and spin off the road. The sheer weight of freezing rain is destructive and is often enough to pull down tree branches and power lines. Freezing rain develops when the lower-level wedge of cold air is so thin that the falling melting snowflake or raindrop doesn't have time to re-freeze before hitting the surface; instead it refreezes when it hits the ground.

Although some people use the terms "sleet" and "freezing rain" interchangeably, they are different phenomena. Because sleet forms before reaching the surface, it bounces off the surface and produces relatively little damage. It is easy to tell when sleet is falling because of the bouncing of the tiny pellets and the sound they make when they hit a windshield or the ground. Motorists can adjust their speed accordingly when they see the sleet hitting their windshields. Because freezing rain does not freeze until impact, it does not bounce; instead, it sticks to whatever it hits. Because freezing rain freezes on impact with the surface, to an observer it often appears the same as rain. But actually motorists are driving on a sheet of ice, even though they are all too often unaware of that until it is too late.

Snow in Florida

Snow rarely falls in Florida, but when it does it is quite a sight. Figure 4.5a is a photograph of the snow that blanketed the town of Live Oak in north Florida during the bitter Arctic cold outbreak in 1899. North Florida was once again covered in snow when a storm hit in 1958. Figure 4.5b shows a home and an automobile in Tallahassee covered in snow during the 1958 snowstorm.

The last significant snowfall in Florida was during 1977, and it made the record books. While people from the North may scoff at the amount of snow that fell over Florida, it was an amazing occurrence. In mid-January of that year, an Arctic cold front passed through Florida, dropping temperatures below freezing. A second cold front came in on the heels of the first, but that front had some accumulated moisture that fell as snow, and it stuck. Snow was reported everywhere across northern Florida. Most of central Florida reported at least a trace of snow, but Tampa and Plant City had 0.2 inches of snow that covered automobiles for a few hours, long enough for photographers to snap some iconic

Figure 4.5. A (top), Snow blanketed the town of Live Oak and much of north Florida during the bitter outbreak of Arctic cold in February 1899; B (bottom), Snow covering a home and automobile in Tallahassee during the 1958 snowstorm. Source: Florida State Archives.

pictures. Snow was reported all the way down to Homestead and even over the Bahamas.

Compared to the rest of the United States, the frequency of snow, sleet, and freezing rain in Florida is, not surprisingly, very low. According to the National Weather Service, parts of the midwestern and northeastern states experience freezing rain an average of about five days per year (Changnon and Karl 2003). The number of days that Florida experiences winter weather is lower than anywhere else in the country. It is nice that residents of the Sunshine State can rest easier than most other Americans regarding the threat of at least one weather-related hazard.

5

Thunderstorms, Rainfall, Hail, and Lightning

The Sunshine State is well known for almost daily lightning-filled thunderstorms in the summertime. These towering storms are visually spectacular and at times scary as the winds blow, rain pelts down, hail falls, and the deafening flash-bang of lightning sounds off. Lightning is a major threat to life in Florida. As clouds grow and skies darken, it is best to seek shelter before the first bolt of lightning and the accompanying shockwave known as thunder arrive. These thunderstorms occur so frequently because Florida has a unique geography; few locations on earth have a substantial peninsula that extends into warm waters at a fairly low latitude. This shape and location leads to the unique development of rumbling thunderstorms and the resulting rainfall.

Precipitation is one of the most fundamental aspects of weather and climate and is dependent on rising air and moisture, which involves a series of complex processes. Ask any meteorologist in Florida and they will tell you that predicting the timing, location, and amount of rain is a humbling experience. Florida has a distinct dry season that lasts from fall into spring. Instead of the almost daily rainfall that Floridians experience during the summer, in the winter, precipitation occurs roughly weekly with the passage of cold and warm fronts. This occurs mostly as rain, but sometimes snow, sleet, and freezing rain falls in northern Florida; it rarely intrudes as far as central Florida.

Thunderstorm Ingredients

The ingredients necessary to create rain are quite basic: humid air and a mechanism, termed "lift," that causes the moist air to rise in the atmosphere. Some places on earth have humid air but no mechanism to allow it to rise. An example is coastal Saudi Arabia and the Persian Gulf region, which is nearly surrounded by water. If you exited an airplane onto the runway on a typical morning there, you would immediately be greeted by the feeling of humid air and your eyeglasses might fog up. But sinking air associated with many regions in the subtropics prohibits this humid air from rising in the atmosphere. This is why places in the Arabian Peninsula and the Persian Gulf region are deserts or semideserts. Other places on earth have mechanisms that allow air to rise vertically, but moisture from the oceans cannot reach the area, either because mountain ranges block the humid air from entering (as in the case for Nevada and Arizona) or because of sheer distance from the moisture source.

For much of the year, Florida has abundant atmospheric moisture. Humid air arrives from the local winds that flow across the Atlantic Ocean or the Gulf of Mexico; moisture blows in from almost any direction except north. The wind direction often determines the character of the precipitation and where it is most likely to fall. Recall from chapter 3 that Florida's prevailing winds are not like those of the rest of the conterminous United States, where prevailing winds typically come from the west. Most of the time, peninsular Florida's prevailing winds are linked to the trade winds, which are from a general easterly direction. During the winter, the winds change direction from northwest after a cold front passes, to easterly as high pressure settles in, and back to southwest as the next cold front approaches. In the summer months, winds are generally from the south, southeast, and southwest.

For example, moist air is typical in Miami in nearly every month of the year (National Climatic Data Center 1998; NOAA 2016). Moist winds come from the southeast every month except January, when they come from the north-northwest. Farther north, the winds and resulting moisture become more variable. Daytona Beach has winds that average from the east to northeast, bringing moist Atlantic air during about half the year—February, May, June, and August through October. Jacksonville, farther to the north, has winds from the easterly half of the compass only in September and October (from the northeast in both months). But even in Daytona Beach, Jacksonville, and elsewhere, south to southwesterly winds can bring moist air from the Caribbean Sea and Gulf of Mexico.

Despite Florida's subtropical location, where air often sinks, the global-scale circulation of air gives Florida a mechanism that allows this air to rise. The subtropical anticyclones described in chapter 3 are associated with sinking air, but the most intense sinking tends to be on the eastern sides of the anticyclone; the western side permits more air to rise, and it rises to greater heights. As we have seen, the subtropical anticyclone that affects Florida most directly is the Bermuda-Azores high. Because Florida is located to the west of this high, air can rise more easily over Florida and the rest of the southeastern United States than it can over western subtropical Africa, home to the world's largest desert, the Sahara.

Free Convection

Warm-season thunderstorms are a common occurrence in Florida. They occur almost daily somewhere in the state in the afternoons and evenings from June into September. The patchwork of land cover, which features urban built-up land adjacent to inland lakes, farmland, forested areas, and ocean, promotes the differential heating of the land surface. This is because such a variety of surface cover types absorb solar (shortwave) energy at different rates. When a small area of the earth's surface is heated more than adjacent areas are, the heated air rises relative to the somewhat cooler air around it. For example, on a warm summer afternoon, air just above a blacktop parking lot will be hotter than the surrounding air and therefore will be likely to rise freely. This is because areas that are heated more intensely tend to support rising motion. In Florida, abundant moisture is typically available; the dewpoint temperature is often only slightly less than the air temperature. Thus, the air does not need to rise to great heights before it cools enough to begin condensation.

The rapid rising of air that lifts moisture to great heights happens in several ways. Temperatures cool with height—sometimes faster and sometimes slower. Generally, temperatures get colder with increasing elevation, particularly above 10,000 feet. That's why, when you fly out of Orlando, Miami, or Tampa in the summer, the ground temperature could be 90°F, but by the time you are cruising at 30,000 feet, it may be -30°F. As warm air rises, it becomes less dense and more buoyant and it cools. If enough moisture is present in the rising air, it will condense. Tall, narrow clouds known as cumuliform clouds begin to form. Imagine a growing puffy cloud as a hot-air balloon. If the hot-air balloon is rising into colder air, the balloon will continue to rise. Similarly, as the warm air rises into colder air, the warm air will continue to rise, and as the moisture condenses, the cloud can grow to great heights. Outside the rising areas, the air is often sinking. When heated air causes this rising-sinking process without the assistance of external forces, the process is called free (or natural) convection. The colder the surrounding atmosphere, the faster the warm air rises and the more vigorous the convection. Because of the localized nature of free convection, the resulting precipitation tends to be isolated and localized. Free convection is more likely to happen in the summer during daylight hours and in warm climates.

The vertical pileup of water in clouds can happen so quickly that water can exist in all three phases of matter within the same cloud—as a gas (water vapor), as a liquid (water droplets), and as a solid (ice). The base of a Florida cloud is generally composed solely of liquid water droplets and water vapor. The cloud may grow above the freezing level to where the upper part of the cloud contains frozen water particles or ice. These clouds are described as having a glaciated cloud top. Even in the summer, ice exists in Florida's clouds, but the ice crystals are concentrated in the highest parts of the cumuliform clouds, where the water vapor has risen into the coldest air. If the rising motion is fast enough, some of the water will remain in liquid form, even though it is at a height above the freezing level. This supercooled water is responsible for hail development. When the cumuliform cloud has sufficient moisture and lift to grow to such heights that ice crystals are abundant at its top and liquid

Figure 5.1. Downburst at Satellite Beach on the east coast of Florida. Credit: Charles H. Paxton.

water is abundant at its base, it is called a cumulonimbus cloud. When so much water in the form of rain and ice has been suspended in the cloud and the updraft weakens, it all comes down, sometimes with great force in a downburst (figure 5.1).

As was described in chapter 4, a special type of graph called a Skew-T allows forecasters to determine how the temperatures, moisture, instability, and winds are distributed through the atmosphere. The data on Skew-T graphs are collected from radiosondes that are carried aloft by large weather balloons.

Forced Convection

Often the atmosphere is cold enough aloft to sustain rapid rising motion but not quite warm enough near the surface to initiate rising motion; it needs a nudge. This nudge, or sometimes shove, is called forced convection. It is initiated by winds that encounter obstacles such as mountains or denser masses of air. When humid air that is being blown by winds encounters a mountain, it is forced to rise. As it rises, it cools and its moisture may condense. This is why the tops of mountains are often shrouded in fog or clouds. The nearly flat topography of Florida cannot produce this orographic lifting. However, even though Florida doesn't have mountains or other topographic features that would cause lift, forced convection is responsible for much of Florida's precipitation. The three primary mechanisms that force convection are fronts, sea breezes, and thunderstorm outflow boundaries.

Fronts are the primary mechanism that create forced convection in most of the United States. During the cool season, forced convection in Florida is usually caused by cold fronts and warm fronts. A front, as described in chapter 4, is simply a linear zone where two air masses with different temperatures meet. Wherever the leading edge of a large mass of cold, polar air encounters much warmer, tropical air, the warmer air is wedged upward over the colder air, producing lift and forced convection. If the air is lifted high enough, the rising air will cool enough for its moisture to condense and cloud formation will begin. Precipitation may follow if there is enough moisture in the air and the rising motion persists. Stronger cold fronts are capable of generating fast-moving lines of thunderstorms, known as squall lines, that often produce damaging wind and tornadoes.

Sometimes forced convection occurs along zones where the warm tropical air is pushing the cold polar air back northward along the polar front. This is a fairly common occurrence in Florida in the late fall, the winter, and the early spring because of the clockwise circulation around the Bermuda-Azores high. In these warm fronts, the warm tropical air rises over the colder air mass, just as is the case with a cold front. Compared to precipitation along cold fronts, precipitation associated with warm fronts tends to last longer and be more widespread, but warm fronts are weaker in intensity. This is because the warm tropical air gains northward momentum from the circulation of the Bermuda-Azores high, allowing the moisture to be slanted northward as it gradually rises over the colder air; instead of existing in a more vertically oriented cloud column along a cold front.

Sea breezes are also important in initiating and focusing convection over Florida. Much like a cold front, sea breezes form a steep wedge that lifts the hot and humid air located inland and kick-starts the convection.

Over the Florida Peninsula, double sea breezes from the Gulf of Mexico and Atlantic Ocean often form and progress inland, initiating rain showers and thunderstorms. When the two sea breezes merge, low-level forcing increases the lift and moisture and the resulting thunderstorms often become quite violent. If the wind flowing over the state is weak, the two sea breezes converge over the spine of the Florida Peninsula. If the steering winds flowing over the state are mainly from the east or the west, the two sea breezes meet closer to the opposite coast, causing storms to intensify there.

Thunderstorms create hail that often melts into rain as it falls toward the ground, cooling the surrounding air, and causing air from the middle of the storm to plunge downward and then outward along the ground away from the storm. This cool outrush of air, called an outflow boundary, feels refreshing and is similar to a small, short-lived cold front. The cool air is denser and acts like a wedge to lift the surrounding warm, moist air; this process initiates more convection. Similar to what happens when two sea breezes collide, when two outflow boundaries or an outflow boundary and a sea breeze collide, convection can be amplified and stronger thunderstorms can result.

Seasonality of Florida Rainfall

How much rain falls in Florida? The statewide average annual precipitation of about 54.5 inches is among the highest of any state. Only Hawaii (63.7 inches), Louisiana (60.1 inches), Mississippi (59.0 inches), and Alabama (58.3 inches) have higher average totals. The opposite ends of the state, northwest and southeast Florida, receive the most precipitation, each averaging over 60 inches annually (map 5.1).

The Florida rainy season starts around the beginning of June (and even earlier in some years) and lasts into October. During the rainy season, southern Florida receives 70 percent of its annual rainfall and northern Florida receives just less than 60 percent. The least amount of annual rainfall, less than 45 inches per year, is over north Florida between Tallahassee and Jacksonville. As a broad generalization, the peak rainfall in northern Florida happens during the cool season and is mostly associated with fronts. In contrast, the peak rainfall in the peninsula occurs

Map 5.1. Average annual precipitation totals across Florida, 1981–2010.

during the warm months due to sea breezes and tropical activity. More specifically, the Panhandle from Pensacola to Tallahassee has a bimodal rainfall distribution with peaks in March and July.

Map 5.2 shows that for January, February, and March, the average monthly rainfall for Florida is greatest over the Panhandle, which receives an average of over 5 inches in each of these months. During those months about half as much rain falls over southern Florida. By May, the totals have decreased over the Panhandle and increased over southeast Florida. By June, rainfall amounts almost double everywhere as the rainy season begins. Map 5.2 also shows that peak average monthly rainfall in the summer months through September is over 8 inches over southern

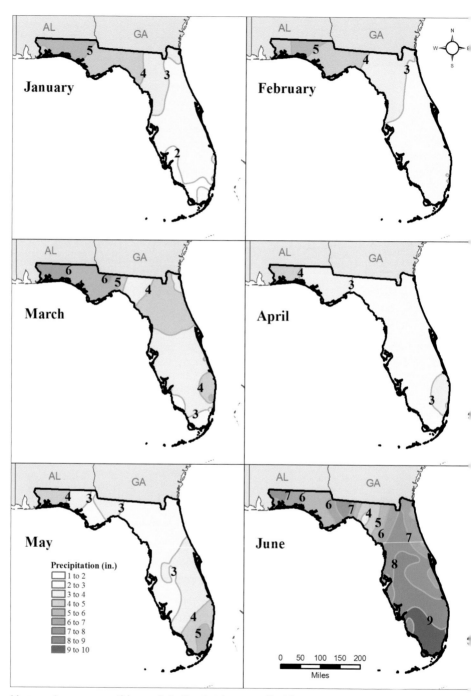

Map 5.2. Average monthly precipitation totals across Florida, 1981–2010.

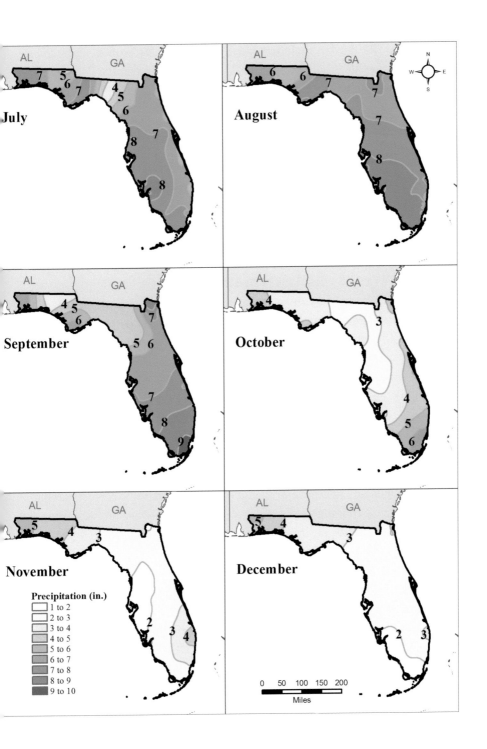

July

August

September

October

November

December

Precipitation (in.)
- 1 to 2
- 2 to 3
- 3 to 4
- 4 to 5
- 5 to 6
- 6 to 7
- 7 to 8
- 8 to 9
- 9 to 10

0   50   100   150   200
Miles

Florida. The rainfall dwindles from October to December, when the greatest average monthly rainfall is around 5 inches over the Panhandle and about half as much over southern Florida.

Warm-Season Thunderstorms

Warm-season thunderstorms usually occur between the months of May and September in Florida. At that time of year, the semi-permanent Bermuda-Azores high pressure system extends westward to influence Florida more directly, creating relatively light winds. The daytime heating of the peninsula's interior begins to set up the sea breezes from both coasts. Moist marine air is drawn into the interior behind the sea breeze and acts as a wedge to lift the heated inland air. As the hot, moist surface air rises, it cools and condenses, intensifying the vertical movement of the interior air column, which in turn intensifies the sea breeze, resulting in the classic Florida afternoon thunderstorm. Latent heat is released to the atmosphere as the hot inland air is lifted, rises, and condenses. This typical afternoon thunderstorm formation is a product of free or natural convection. By "typical" we mean that the thunderstorm is not associated with a larger, more organized storm system. The typical afternoon thunderstorm, sometimes known as an air mass thunderstorm, is its own entity; it usually forms and dies within an hour or two. These thunderstorms seldom cause widespread severe damage but often cause localized deadly lightning strikes, gusty winds, and traffic accidents. The most intense thunderstorms occur where the west-coast and east-coast sea breezes collide, creating extra lifting motion.

The Effect of the Coastline on Thunderstorms

The shape of a coastline also has a major impact on the timing and location of precipitation. The coastline shape influences the direction of the sea breeze's propagation inland. Sea breezes generally blow perpendicular to the coastline and progress inland. The cloud band associated with a sea breeze aligns parallel to the coast. As was shown in chapter 3, this allows for the sea breeze line to develop in convex, concave, or straight-line forms. For example, where a coastline extends seaward, as is the case near Apalachicola, Cape Canaveral, areas of Tampa Bay, and

Fort Myers, the sea breezes push inland at an angle and converge. The opposite is true when the coastline has a broad indentation. There, sea breezes form an arc shape and the clouds spread apart as they move inland. The shape of the land around Tampa Bay focuses the sea-breeze convergence and generates more frequent and persistent thunderstorms; this area of the state averages 80 to 100 thunderstorm days each year. This is a greater number of annual thunderstorm days than in any other part of the United States. The opposite is true along a concave coastline, which tends to spread out the sea-breeze front. Other properties of the coastline and the shape of land influence sea breezes. Researchers Xian and Pielke (1991) determined that the width of a peninsula can strongly influence the development of a sea breeze and noted that roughly 100 miles, the width of Florida, is ideal for the development of convection and the strongest vertical velocities within thunderstorms as sea breezes merge. This is another way that Florida's geography is unique.

The Timing of Thunderstorms

Depending on daily weather conditions, the location and timing of afternoon thunderstorms can vary greatly. Steering winds are usually dictated by the semi-permanent Bermuda-Azores high pressure, which extends east-west somewhere across Florida. The winds that steer thunderstorms are between 5,000 to 10,000 feet above sea level and sometimes higher. When the high pressure situated across Florida is weak, the winds are weak. Under weak wind flow, sea breezes develop along the coasts during the late morning and meander slowly inland, meeting somewhere in the middle of the peninsula. Weak wind flow days often produce the greatest localized rainfall amounts because the storms just move slowly and dump rain.

If a high-pressure ridge is situated over north Florida or Georgia, winds will be from the east or southeast over the peninsula, resulting in an extra push of the east-coast sea breeze inland. Under this flow regime on the west coast, the east to southeast winds push against the west coast sea breeze and slow the inland movement. In this case, the east- and west-coast sea breezes will meet close to the west coast of the peninsula, and that is where the most intense thunderstorms will be located. When cold fronts push into the southeastern United States during the summer, the Bermuda-Azores high that often extends across central Florida is pushed southward and much of the state is under a southwesterly wind flow. In this scenario, the sea breeze and resulting thunderstorms develop early in the day along the west coast of Florida and propagate inland, moving toward the east coast. Along the east coast, the development of the sea breeze and its movement toward the inland areas is slowed by the opposing wind. In this scenario, the sea breezes collide on the eastern side of the peninsula, producing strong thunderstorms. Only one sea breeze forms along the Panhandle and it is also steered by the winds. Southerly winds across the Panhandle will push the sea breeze farther inland, while easterly or westerly winds will result in less movement inland.

Hail in Florida

Hail is one of the mysterious types of precipitation that occurs in Florida. Imagine pieces of ice falling from the sky on an otherwise warm day. It is not a welcome sight. Hailstorms begin with a few small stones falling from the sky but as the hail grows larger, it is noisy and scary as it bashes automobiles and rooftops with force. Hail in Florida can grow to the size of baseballs or softballs in the clouds high above the earth, but it usually melts to a size that ranges from the diameter of a pea to the diameter of a golf ball before hitting the ground. Hail the size of a golf ball can inflict serious damage to whatever it strikes. This is one of the most destructive forms of precipitation because it destroys crops and shreds leaves off trees. Larger hail falls at speeds over 70 mph and can break car windows, dent metal, damage roofs, and tear through screen enclosures. Although large hail has killed people in other places, no records exist of a death caused by hail in Florida, although it can certainly batter and bruise anyone caught without cover.

Hail Formation

Hail forms within a cumulonimbus cloud updraft. Extreme updrafts create a condition in which liquid water is supercooled by being pulled rapidly upward, where temperatures are below freezing. The supercooled water then freezes and the nearby water vapor spontaneously bypasses

Figure 5.2. Large hail with concentric growth rings. Source: National Weather Service and NOAA (2006).

the liquid phase and becomes ice when it touches the frozen water droplets. This causes the tiny pellet to grow. Hail continues to be suspended within the updraft as it continues to grow from more of this supercooled water deposition. Hailstones may be comprised of many layers of frozen water and may aggregate with other hailstones (figure 5.2). Eventually the hailstones either grow too large to be supported by the updraft or the updraft weakens and the hail begins to fall toward the ground (figure 5.3).

A three-dimensional radar image of a hailstorm that produced grapefruit-sized hail in March 1996 (figure 5.4) shows several layers of storm intensity. The orange colors represent moderate rain, the magenta color indicates a mix of rain and hail, and the white core area represents large hail. The magenta color is adjacent to what is known as a bounded weak echo region. Weak echo regions, which are common in radar depictions of hail-producing thunderstorms, indicate the thunderstorm updraft region. During large hail events, the ingredients include atmospheric moisture, strong instability (the likelihood that enough warm air will rise freely to produce hail), cold temperatures in the middle part of the

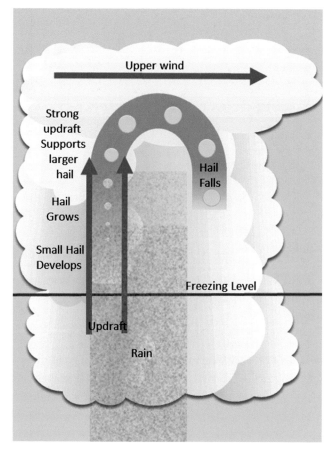

Figure 5.3.
Hail growth.

atmosphere, and wind shear. The latter increases the ability of the up-draft/downdraft system to remain intact and become a supercell.

Extreme Hail

Imagine hail the size of grapefruit falling out of the sky and smashing to the ground. This is what happened on March 30, 1996, when hail 4.5 inches in diameter fell in Florida. This is the largest hail known to have fallen in the state. Hail of that size had been recorded only two other times in Florida. The long-lived hailstorm was tracked by radar as it dropped nickel-sized hail in the Tampa Bay area and moved eastward, growing in strength. The hail that bombarded the central Florida citrus-growing region was produced by a supercell thunderstorm that grew in

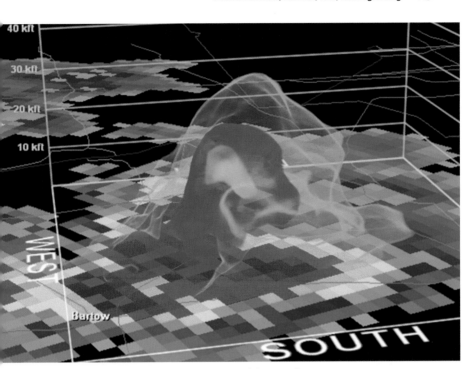

Figure 5.4. Three-dimensional radar image of the March 30, 1996, storm over central Florida that produced grapefruit-sized hail. The outer shaded area depicts moderate rain, the darker interior color indicates rain and hail, and the white shaded area is large hail. Source: Unpublished information from local National Weather Service office.

intensity as it moved across Polk County. Hail that was mostly the size of baseballs (although some was the size of softballs) destroyed roofs and knocked out windows in nearly 600 homes and cracked the windshields of around 3,000 cars in Lake Wales. Most of the damage was done to the north side of homes and buildings. The local fire station reported hail covering the ground nearly six inches deep during the peak of the hailstorm. The estimated cost of damage to homes, businesses, and automobiles was nearly $24 million.

Florida's record hail, though, was only about half as large as of the largest hailstone on record in the United States. In 2010, hail that was 8.0 inches in diameter and weighed 1.94 pounds fell near Vivian, South Dakota (figure 5.5).

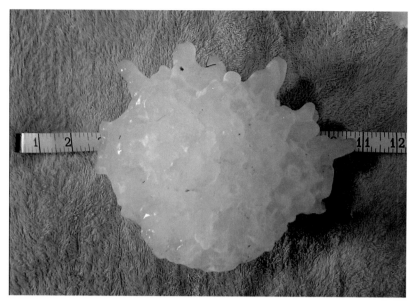

Figure 5.5. Eight-inch-wide hail that fell near Vivian, South Dakota, in 2010. Source: NOAA photo library. Source: National Weather Service and NOAA (2010).

Sometimes the hail isn't large but it is plentiful. During a hailstorm on March 6, 1992, in Longwood, Florida, drivers found their cars parked in what looked like a snow bank; it was 2 1/2 feet of small hail. A photo of the event made the cover of the NOAA publication *Storm Data* (figure 5.6). A similar hailstorm dumped several feet of smaller-sized hail near Zephyrhills, Florida, on January 29, 1997.

Frequency of Hail across the United States

Map 5.3 shows the average number of hail days per year across the United States. Some locations in the central United States have over 25 days of hail a year. The National Weather Service collects data on hailstorms from citizen reports; this is why statistics are less complete for less-populated areas. The NOAA Storm Prediction Center indicates that the Great Plains—eastern Colorado, western Texas, Oklahoma, Kansas, and Nebraska—experience an average of more than 10 days of severe hail each year, more than anywhere else in the United States.

Figure 5.6. A hailstorm on March 6, 1992, in Longwood, Florida, dumped 2.5 feet of small hail on the ground. Source: NOAA Storm Events Database.

The Bermuda-Azores high, and sometimes transient high-pressure areas, transport moisture in a southerly flow (south to north) into the Great Plains. Whenever a cold front approaches the Plains with enough moisture, wind shear, and strong rising motion (instability), hail is a possibility. The instability is linked to sufficient surface heating to initiate cloud growth and cold temperatures in the mid-levels of the atmosphere, which leads to stronger uplift. This part of the country is more likely to have supercell thunderstorms with strong updrafts capable of supporting large hail growth.

In the western Great Plains, hail occurs most often in the spring because during that time of year very cold air behind a cold front can interact with much warmer and more humid air ahead of it. This creates a situation with strong instability that allows strong updrafts to extend to great heights. In any given April, a resident of northwestern Texas, western Oklahoma, Kansas, Nebraska, or eastern Colorado is likely to see hail at least once, while a Floridian is much less likely to do so. The

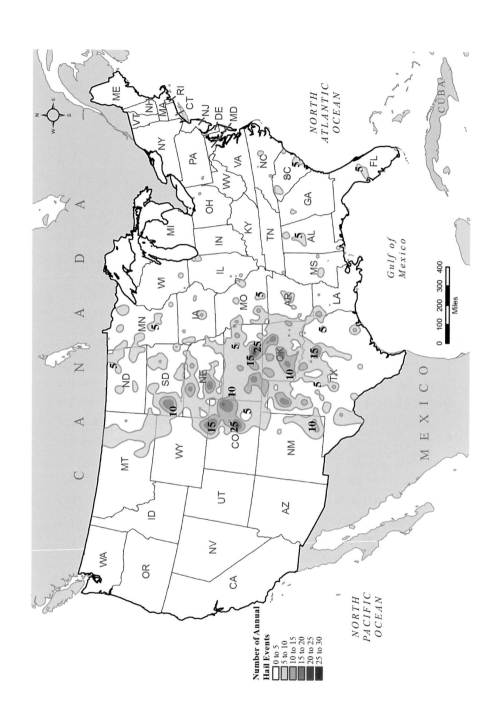

western Great Plains probably experiences hail as often as anyplace else on the planet, but the lack of global hail records makes this a difficult assertion to prove. Farther north, in the Dakotas and Minnesota and into central Canada, hail is most common in summer, because the interface of cold and warm air retreats northward over that area in summer. In any given July, a resident of the Dakotas or Minnesota is likely to see hail at least once, while a Floridian is far less likely to see hail in that month.

Hail rarely falls over the western United States, where moisture is usually too limited to generate clouds with sufficient vertical extent to support the organized system of updrafts and downdrafts required. Along the West Coast of the United States, moisture is sufficient at times, but the updrafts are not vigorous enough because the land does not warm enough to create sufficient free convection.

Frequency of Hail in Florida

Compared to other parts of the United States, Florida has only modest hail frequency. Most of Florida receives hail only once or twice a year, and it is usually small. The central peninsula portion of Florida, along the I-4 corridor, has 5 to 10 days of hail per year, the highest hail frequency in the state. This is mainly due to the combination of colder temperatures aloft than we might see in south Florida and strong sea-breeze collisions. Both of these result in the strong updrafts necessary for hail formation. Another reason for the higher frequency of reported hail across the central peninsula, from Tampa to Orlando, is that the population is greater and reports of hail are more likely.

Several factors limit the frequency and size of hail over Florida. First, during the summer thunderstorm season, the atmosphere is often too warm across a very deep layer of the atmosphere for hail to reach the ground without melting first. During the summer, freezing levels in Florida are typically above 13,000 feet with plenty of warm air beneath to melt hail as it falls. During the cool season, although wind shear sometimes accompanies and organizes thunderstorms into supercells with strong updrafts, other factors such as moisture and instability are often missing and thunderstorms fail to develop. During the predominant summer thunderstorm season, when moisture and instability are abundant, the wind shear is too weak to organize and sustain rotating cumulonimbus supercell clouds for periods of an hour or more, which lowers the chances that hail will develop and strike the ground.

Florida Hail Seasons

The largest hailstones in Florida fall during the cool season, from December through April; March is the peak month for large hail. This seasonality contrasts somewhat with most of the rest of the United States. The largest hail is associated with strong low-pressure systems that extend over the Gulf of Mexico and the southeastern states. These systems pull warm, moist air from the Caribbean region northward over Florida. In the middle to upper levels of the troposphere, the air is colder than usual and the winds are stronger and increase with height. This is the formula for supercell thunderstorms, which rotate, persist, have strong updrafts and are capable of keeping large hailstones in the air as they continue to grow.

However, hail is most frequent in Florida during the warm season, from May through September. The months of May, June, and July account for 60 percent of hail occurrences in Florida; it occurs most often in June. Warm-season hail stems from frequent thunderstorm cells that are reasonably well organized but generally too poorly organized and too short-lived to grow large hail. Thus, the hail that forms experiences significant melting as it falls to the ground.

Lightning

Although over 70 percent of the earth is covered by ocean, only 10 percent of lightning strikes over the vast ocean waters. The other 90 percent of global lightning strikes the continents, where the heated land provides more energy for convection. Lightning is more widespread near the equator, where convection is more frequent and vigorous, and it generally decreases toward the poles.

Despite Florida's undisputed claim that it is the lightning capital of the country, the state pales in comparison to other lightning hot spots. Figure 5.7 shows the global annual frequency of lightning flashes from 1998 to 2013. It was created with data from the Lightning Imaging Sensor

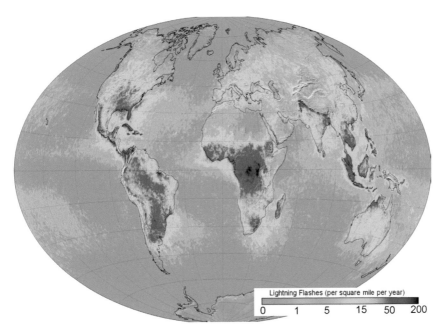

Figure 5.7. Global lightning frequency. Source: NOAA Storm Events Database.

(LIS) aboard the Tropical Rainfall Measuring Mission (TRMM) and the Optical Transient Detector (OTD) aboard the Microlab-1 spacecraft. Areas with the largest number of lightning strikes per year are in black.

Deep in the heart of the African continent, lightning strikes a broad area in the Democratic Republic of the Congo more than 250 days per year. It strikes about 400 times per square mile each year in this area. However, in 2016, Rachel Albrecht and other researchers deemed Lake Maracaibo Venezuela, the largest lake in South America, as the new lightning top spot (Albrecht et al. 2016). The lake, located in the Andes Mountains, is where lightning strikes an average of 297 nights per year. As the sun sets and the surrounding mountains cool, downslope winds converge over the warm waters of Lake Maracaibo, creating convective storms with vivid lightning flashes.

Lightning in Florida

Lightning strikes on 70 to 100 days each year virtually everywhere in Florida. Studies of cloud-to-ground lightning show that the greatest annual number of lightning strikes occur in central Florida along the I-4 corridor. However, during the cool season, from November to February, the number of lightning strikes is greatest over the Florida Panhandle. This is mostly due to thunderstorms that form along vigorous cold fronts over the Panhandle during that time of year. During the transition to the warm season, from March to May, lightning strikes increase over the entire state. During the summer, the greatest amount of lightning activity occurs in the Tampa Bay area. The geography of the region plays a strong role there; the shape of Tampa Bay and the land interface create stronger wind convergence.

As we have seen, different wind flows affect how sea breezes move, where convection forms, and where lightning strikes. Along the Panhandle, changing wind regimes have less impact on lightning than they do over the peninsula, where the east- and west-coast sea breezes collide. Under calm wind regimes, these sea breezes drift inland and meet along the spine of the peninsula. When the flow is from an easterly direction, lightning is concentrated along the west coast. Under westerly flow, lightning is concentrated along the east coast of the peninsula. Additionally, collisions of sea-breeze boundaries lead to more frequent and stronger thunderstorms.

Other lightning hot spots in the state are located where coastal land extends into adjacent water areas. Such areas include Apalachicola, Bradenton, Fort Myers, Palm Beach, and Cape Canaveral. Every square mile of central Florida averages 25 cloud-to-ground lightning strikes per year. Florida also leads the nation in lightning deaths, averaging six per year. Texas follows with half the number of annual deaths.

Lightning Safety

The injuries of lightning survivors can be debilitating, requiring extensive hospitalization and lifelong care. The safest place to be during a thunderstorm is inside a building away from windows and wiring. Metal cars are also a safe location. When an automobile is struck, the lightning moves quickly from the metal body to the ground. If you are in a car during a lightning storm, it is best to be parked because several instances of airbags deploying during a lightning strike have been documented.

Lightning also causes millions of dollars in property damage to buildings, homes, and electronics.

The Development of Lightning

Because of the difference in the physical properties of ice crystals and water droplets, ice and water tend to attract charged particles of opposite polarity. Positively charged particles are attracted to ice crystals, and negatively charged particles are attracted to water droplets. Recall that ice and water can co-occur in clouds. On clear days, negatively charged particles are also attracted at the surface of the earth. Usually, there is no need for an exchange of electrical energy between the base of the cloud and the earth's surface. But when a mature thunderstorm cloud, that has many negatively charged particles around the liquid water at its base, passes over the ground, an opposite positive potential attracts positively charged particles on the earth's surface (figure 5.8).

This positive ground potential follows the cloud like a shadow. The positively charged particles on the surface attempt to get as close as

Figure 5.8. Lightning development. Source: NOAA, adapted by Melissa Metzger.

possible to the negatively charged particles. Under ordinary circumstances, the electric potential cannot equalize because air is a good insulator and does not easily conduct electrical energy. Thus, an extremely high voltage gradient (millions of volts) can exist within the clouds and the ground. The charges are not usually exchanged within the cloud, but when the voltage potential becomes sufficiently high, small segments of energy called stepped leaders travel downward. At the same time, a positively charged stepped leader moves up from the ground. When the two stepped leaders meet, a low resistance path is established and a surge of electrical current moves from the ground to the cloud, creating the electrical discharge—lightning. Slow-motion photography illustrates this fascinating process, which is too fast for the human eye to discern.

Lightning flashes can happen within a single cloud, between two clouds, from cloud to air, and from cloud to ground. According to satellite observations, within-cloud lightning and cloud-to-cloud lightning are the most common types; some estimate that these account for up to 75 percent of all lightning strikes (NASA 2001). That means that only 25 percent of lightning impacts the ground. This is fortunate, because more strikes hitting the earth would cause a much more dangerous planet, since the heat involved in a lightning strike is tremendous. It is estimated that the temperature around a lightning strike is roughly 50,000°F, more than the temperature of the sun's surface several times over.

The intense heat of the electrical energy of lightning causes the air around it to expand outward, away from the strike, creating a shockwave that propagates away the electrical exchange and creates the sound we know as thunder. Because lightning moves at the speed of light and thunder moves at the slower speed of sound, there is a noticeable difference in the timing of our sensation of lightning and thunder. For every five seconds between the time you see a lightning strike and the time you hear the clap of thunder associated with it, the lightning strike is one mile away from you.

Although lightning is bright and hot, its diameter is only about the size of a pencil or even smaller. We know this because when lightning strikes sand or some soils, the intense heat melts the sand around the electrical energy and creates a hardened tube called a fulgurite.

Extreme caution must be taken when lightning is occurring or is expected to occur as clouds build and skies darken. People are often struck by the first bolt of lightning in the area. Lightning can also strike diagonally from a distant storm. Even after the thunderstorm has passed and the lightning appears to be over, the anvil cloud can hold a positive charge and send a deadly bolt to the earth. So be sure that the storm has passed far away before emerging from shelter.

6

- - - - - - - - - - - - -

Tornadoes

Florida might not be the first state that comes to mind when we think of tornadoes, but it experiences more tornadoes per square mile than any other state in the United States. A tornado is a violent, rotating vertical column of air that emerges from a cumulonimbus thunderstorm cloud and creates dangerously strong surface winds in its path. Fortunately, most tornadoes are short-lived and cause only minimal damage. In some years, Florida experiences more tornadoes than any other state. Tornadoes are one of nature's mechanisms for quickly addressing imbalances in the atmosphere. Those imbalances can be linked to thermal instability, which happens when warm air near the surface rapidly rises into cold air aloft, or dynamic instability, when strong or shifting winds through the atmosphere create rising motions that lead to thunderstorm rotation. Sometimes tornado development is a combination of both thermal and dynamic processes. When this happens, the tornadoes that form are often stronger and longer lived than most tornadoes.

The vortex of a tornado rapidly pulls in air and sucks up objects into the swirl. Though the damage tornadoes cause is more localized than in most other types of violent storms, tornadoes produce the strongest winds and the most gruesome path of destruction of any type of storm. The strong winds are loud, like a whistle, especially when flowing through trees and past buildings. Tornado survivors often report that they heard the sound of a freight train or jet engine just before the tornado hit; these sounds are sometimes the first signal that a tornado is

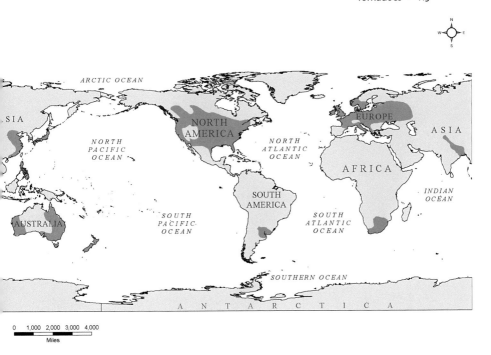

Map 6.1. Global tornado development areas.

approaching. In this chapter, we take a close look at tornadoes and other similar phenomena.

Where Tornadoes Occur

Some areas of the world are tornado hot spots (map 6.1). Argentina, Uruguay, and a sliver of Brazil are likely to experience tornadoes. The Atlantic Ocean and the Mediterranean Sea can be sources of moist air, in Europe and western Russia. Depending on location and wind flow, these areas of the planet are not immune to tornadoes when conditions are right. South Africa is another tornado hot spot. In Asia, tornadoes are more common in flat lands adjacent to the Himalaya, along the coast of China, and over South Korea, Japan, and the Philippines. In the area known as Oceania, parts of Australia and New Zealand regularly experience tornadoes and waterspouts, which are tornado-like vortices that form over water.

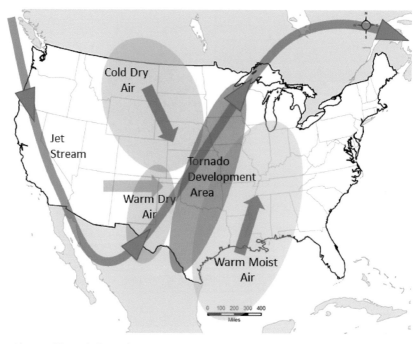

Map 6.2. Tornado ingredients.

But of all the countries in the world, the United States experiences the most tornadoes. Warm, moist air from the Gulf of Mexico meets the much colder air sliding south from the polar regions (map 6.2). At the same time, dry air filters in from the desert Southwest. Above these interactions is a strong jet stream that increases wind speed with height, increasing wind shear. These disparities in air masses in the atmosphere lead to a process that spawns tornadoes. This pattern occurs more frequently across the Great Plains and the southern tier of states, but it can shift, bringing tornadoes to other parts of the country.

In the United States, two regions have a high frequency of tornadoes. The term Tornado Alley has been used for years to describe the swath of land from Texas north into Iowa, where frequent, powerful tornadoes are more prevalent than elsewhere in the country. The second major swath of tornadoes occur in Dixie Alley, an area that spreads across the adjoining Gulf Coast and the southeastern states from Louisiana up into the Carolinas and includes parts of Florida.

Florida Tornadoes

Florida's almost daily summer thunderstorms and occasional cool-season thunderstorms lead to the large number of tornadoes that develop. Florida has three tornado seasons: the summer, the season of tropically induced tornadoes, and the winter. During the summer, cumulus clouds that rapidly grow into cumulonimbus clouds spawn weak tornadoes and waterspouts. These often develop along the sea breezes and land breezes that move onshore and offshore. Later in the warm season, tropical storms and hurricanes create bands of showers and thunderstorms that generate fast-moving tornadoes that typically develop on the right side of the tropical weather system as you face the direction of movement. The third tornado season is during the cool season. As cold air masses plunge southward, the combination of strong differences in air mass and a strong polar jet stream near Florida can create conditions for severe thunderstorms that develop into tornado producers. This often occurs when strong areas of low pressure develop over the Gulf of Mexico, lifting warm, moist air northward over the state ahead of cold fronts. Florida is one of the few places in the United States where strong tornadoes occur during the winter.

Tornado Classifications

Because of their strong winds and short life spans, tornadoes are elusive and difficult to measure. Standard weather instruments such as anemometers to measure wind speed and barometers to measure air pressure are not easily placed in the path of a tornado. Newer technologies such as mobile Doppler radar systems provide estimates of wind speeds within a tornado, but the opportunities for those close-proximity measurements are rare. This is why tornadoes are most often ranked by the amount of death and destruction they cause. It is often the number of fatalities that brands a tornado as notorious. Some tornadoes have become notorious because of their long duration and the number of miles they traveled while ravaging the landscape.

The Fujita Scale

Dr. Ted Fujita (1920–1998), known as "Mr. Tornado" (figure 6.1), was a professor of atmospheric sciences at the University of Chicago, where he studied damaging convective storms. He produced highly detailed maps of tornado damage and found that storms that spawned tornadoes produced different damage patterns than storms that created a simple downward rush of air. The downrush of air from a collapsing or surging thunderstorm is known as a downburst. Damage from downbursts creates a straight or divergent pattern of debris-blown damage like the outstretched fingers on a hand (figure 6.2a). In contrast, tornadoes create a spiraling, convergent damage pattern in which debris is blown toward the center path of the tornado (figure 6.2b).

In 1971, Fujita created a scale that ranked tornadoes from 0 (weakest) to 5 (strongest) based on the damage they created (Table 6.1). Because it was almost impossible to measure speeds within stronger tornadoes, Fujita used the tornado damage done to trees and buildings to estimate the associated wind speeds. The scale, known as the Fujita scale, had rankings from weak tornadoes (F0), which have winds below 73 mph,

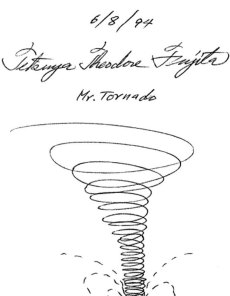

Figure 6.1. Signature of Dr. Ted Fujita, known as "Mr. Tornado."

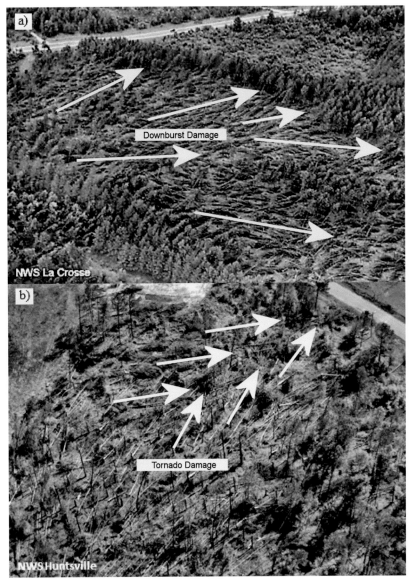

Figure 6.2. A, Divergent damage pattern of a thunderstorm downburst; B, Convergent damage pattern of a tornado. Credit: A, NOAA Storm Events Database; B, National Weather Service Forecast Office, Huntsville, Alabama; National Weather Service (2013).

Table 6.1. The original Fujita scale

Scale	Wind Estimate (mph)	Typical Damage
F0	< 73	Light damage: some damage to chimneys, branches broken off trees, shallow-rooted trees pushed over, signboards damaged
F1	73–112	Moderate damage: surface peeled off roofs, mobile homes pushed off foundations or overturned, moving autos blown off roads
F2	113–157	Considerable damage: roofs torn off frame houses, mobile homes demolished, boxcars overturned, large trees snapped or uprooted, light-object missiles generated, cars lifted off the ground
F3	158–206	Severe damage: roofs and some walls torn off well-constructed houses, trains overturned, most trees in the forest uprooted, heavy cars lifted off the ground and thrown
F4	207–260	Devastating damage: well-constructed houses leveled, structures with weak foundations blown away some distance, cars thrown and large missiles generated
F5	261–318	Incredible damage: strong frame houses leveled off foundations and swept away, automobile-sized missiles projected through the air for more than 100 yards, trees debarked. Incredible phenomena will occur.

to intense tornadoes (F5), which have winds from 261 to 318 mph. He even included a description of a "super tornado," which he termed an "inconceivable tornado" and gave a theoretical ranking of F6. Interestingly, because his scale was based on damage to trees and structures, limitations existed. When a tornado occurred in an area without trees or structures, the highest rating could only be an F0. The scale did not take differences in building construction into account and it tended to overestimate wind speeds in categories above F3.

The Enhanced Fujita Scale

Using modern engineering standards and measurements, a team of scientists and engineers revamped Fujita's scale. This new scale, which came into use in 2007, is called the Enhanced Fujita scale (Table 6.2). The categories remained zero through five but were preceded by "EF" to

Table 6.2. The Enhanced Fujita scale

Scale	Wind Estimate (mph)	Typical Damage
EF0	65–85	Minor or no damage: surface peeled off some roofs, some damage to gutters or siding, branches broken off trees, shallow-rooted trees pushed over. Confirmed tornadoes with no reported damage (i.e., those that remain in open fields) are always rated EF0.
EF1	86–110	Moderate damage: roofs severely stripped, mobile homes overturned or badly damaged, exterior doors blown off, windows and other glass broken
EF2	111–135	Considerable damage: roofs torn off well-constructed houses, foundations of frame homes shifted, mobile homes completely destroyed, large trees snapped or uprooted, light-object missiles generated, cars lifted off ground
EF3	136–165	Severe damage: entire stories of well-constructed houses destroyed, severe damage to large buildings such as shopping malls, trains overturned, trees debarked, heavy cars lifted off the ground and thrown, structures with weak foundations damaged badly
EF4	166–200	Extreme damage: entire well-constructed frame houses completely leveled, cars and other large objects thrown, and small missiles generated
EF5	>200	Total destruction of buildings: strong-framed, well-built houses leveled off foundations and swept away; steel-reinforced concrete structures damaged critically; tall buildings collapse or have severe structural deformations; some cars, trucks, and train cars can be thrown approximately 1 mile

denote the enhanced scale. A primary difference of the new scale is the lower wind speeds in each category above EF1. These lower wind speeds imply that winds cause more damage than was previously realized. It was not until recently that more accurate wind speed estimates from mobile Doppler radar systems could be used to classify tornadoes. The new scale also had better descriptions of damage, with examples and photos that blended well with the original Fujita scale. The EF Scale has 28 damage indicators that describe the estimated degree of damage to buildings and vegetation for specified categories of wind speed.

Tornado Numbers across the United States

What states have the most tornadoes? The National Severe Storms Laboratory in Norman, Oklahoma, maintains the most complete and current set of tornado statistics for the United States. In the period 1981–2010, Florida was in fourth place, averaging 61 tornadoes per year (map 6.3). In first place was Texas (156), which has a large land area, followed by Kansas (82) and Oklahoma (63). When examining the average annual number of EF0–EF5 tornadoes per 10,000 square miles (map 6.4), Kansas falls into first place and Florida ties for second with Iowa, Oklahoma, and Mississippi. The tornado statistics can be sliced and diced several different ways. Florida is second in the country for the greatest number of tornadoes in a state from 1980 to 1990 and second for the average number of days per year with tornadoes (28). When considering only the strongest tornadoes (F3/EF3 and above), Texas comes to the top of the list, followed by Arkansas, Kansas, and Oklahoma; Florida ranks 26th in this statistical measure (map 6.5).

We can conclude that Florida experiences a higher number of tornadoes than other states but only an average number of intense (F3/EF3 and higher) tornadoes. This is because Florida often lacks the intense atmospheric conditions necessary for the most destructive types of tornadoes. The combination of thermodynamic and dynamic forces invites formation of strong tornadoes. The atmospheric conditions include an intense thermodynamic profile with very cold air overlying very warm air, leading to strong instability. When these conditions are present with strong wind shear, thunderstorm cells may rotate, become more

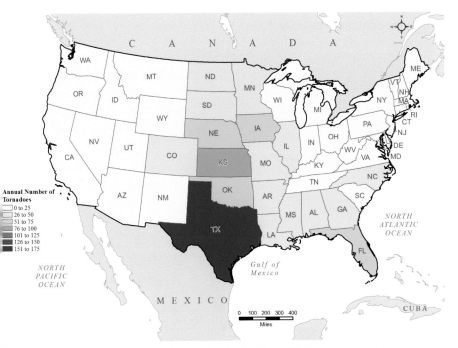

Map 6.3. Average annual number of tornadoes in the United States, 1981–2010. Source: NOAA and National Weather Service Storm Prediction Center (2016a).

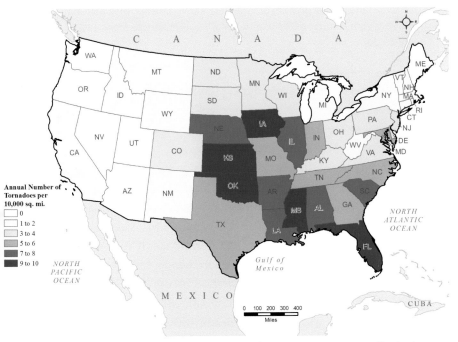

Map 6.4. Average annual number of EF0–EF5 tornadoes per 10,000 square miles in the United States, 1981–2010. Source: NOAA and National Weather Service Storm Prediction Center (2016a).

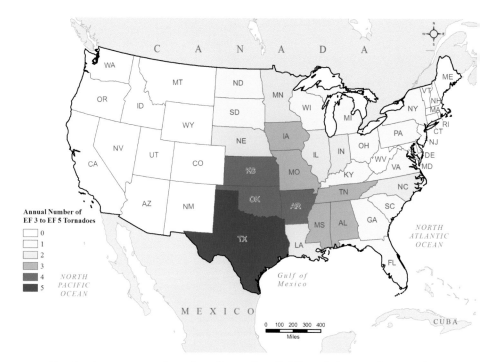

Map 6.5. Average annual number of tornadoes F3/EF3 and above in the United States, 1981–2010. Source: NOAA and National Weather Service Storm Prediction Center (2016a).

organized, and last longer. Wind shear is typically associated with strong upper-level jet streams in association with strong cold fronts. Florida seldom experiences the combination of these types of features that can turn a relatively weak tornado into a strong one.

But despite the relatively unspectacular frequency of intense tornadoes, because of its large population, Florida ranks fifth (behind Alabama, Tennessee, Texas, and Arkansas) in the average annual number of tornado fatalities from 1981 to 2010 (map 6.6). Alabama, Tennessee, Texas, and Arkansas have a high percentage of strong tornadoes but few basements or storm cellars where people can seek refuge. The high population in Florida, especially the high number of mobile home dwellers, makes Floridians particularly vulnerable to fatalities during tornadoes. Map 6.7 shows Florida's deadly tornado tracks from 1990 to 2013. Most of the tornadoes have occurred over the Florida Panhandle

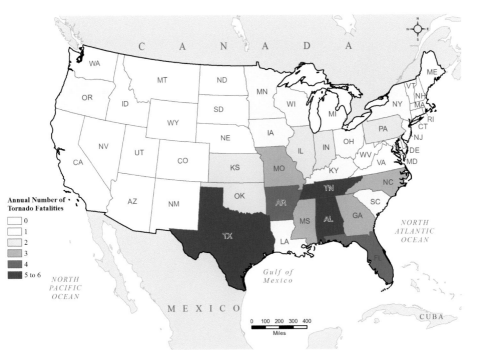

Map 6.6. Average annual number of tornado fatalities in the United States, 1981–2010. Source: NOAA and National Weather Service Storm Prediction Center (2016a).

or the central peninsula region. Most of the deadly tornadoes occurred away from the more densely populated coast, and it is not surprising that no deadly tornadoes were reported over central northern Florida or the Everglades, where the human populations are smaller.

Tornado Development

Tornadoes come in a variety of shapes and sizes and from a variety of cumulus cloud formations (figures 6.3a–c). Some, including most of the storms ranked F0/EF0, are spawned by a single, rapidly growing, nonrotating towering cumulus or cumulonimbus cloud (figure 6.3a). Other tornadoes can develop along multicellular thunderstorm lines that are sometimes called squall lines. Sometimes tornadoes develop on the north or south end of a multicellular bow-shaped echo that is

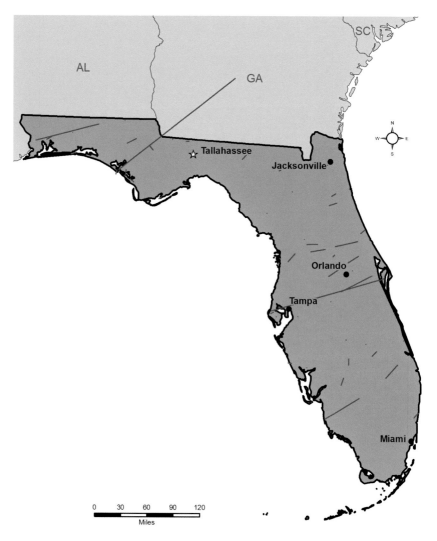

Map 6.7. Tracks of Florida's deadly tornadoes, 1990–2013. Source: NOAA and National Weather Service Storm Prediction Center (2016b).

sometimes known as a derecho, a name derived from the Spanish for "straight ahead" (figure 6.3b). Damage paths from derechos indicate a straight-line pattern of downed trees instead of the inward-swirling pattern caused by tornadoes. Derechos create a distinguishable signature on Doppler radar. They typically produce a broad and persistent moving line of intense straight-line winds that often stretch for 25–100 miles.

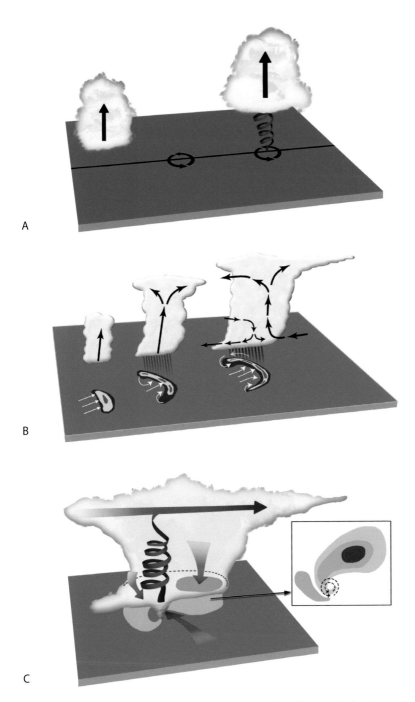

Figure 6.3. *A*, Single-cell tornado development; *B*, Multicell tornado development; *C*, Supercell tornado development. Credit: *A* and *C*, Melissa Metzger; *B*, Anita Marshall.

Their wind speeds can rival those of an EF2 tornado. Some tornadoes, including most of those ranked as F3/EF3 and stronger, form within a single, large thunderstorm called a supercell, which itself is rotating (figure 6.3c). On radar, supercells have a paisley-like appearance: a broad northeast portion and a narrow hooked portion to the southwest. Supercells develop in regions of strong thermal instability and vertical wind shear that cause the thunderstorm updrafts to rotate. The supercell structure feeds converging winds into the strong updraft, and strong upper-level winds efficiently transport residual moisture away from the updraft.

Forecasting Tornadoes

Forecasters scour weather maps, looking both horizontally and vertically, for areas with strong thermal, moisture, and wind discontinuities in order to identify regions where the next tornado outbreak will occur. Sometimes the evidence is strong and the forecasters have great confidence in their predictions. Sometimes only a few pieces of the puzzle fall into place and the threat outlook for tornadoes is placed in a lower category. The National Weather Service Storm Prediction Center in Norman, Oklahoma, issues tornado outlooks for regions of the nation up to several days in advance. The center issues tornado watches covering large segments of the county hours in advance of tornadoes it anticipates. Even with the most effective weather forecast tools, meteorologists can usually predict tornadoes for a specific town or city only up to 20 minutes ahead of time. Since the 1990s, wind velocity measurements using Doppler radar have helped increase the lead time of warnings. According to the National Weather Service, the average lead time across the United States for tornado warnings is 13 minutes (NOAA National Severe Storms Laboratory 2016). The next great breakthrough in tornado warning meteorology will likely be high-resolution computer models that can forecast the timing and location of tornado development hours or even days in advance.

Hurricane-Induced Tornadoes

Tornadoes are common within tropical cyclones and can form when some of the normal conditions for their development are absent. Tornadoes induced by tropical cyclones tend to occur most frequently in the outer convective bands of the right forward quadrant of the storm—the side that experiences the most intense storm impacts in general. The wind fields associated with stronger tropical systems tend to move these tornadoes at fast speeds (30–60 mph). Unlike tornadoes that occur in association with cold fronts, squall lines, or supercells, tornadoes induced by tropical cyclones tend to move from southeast to northwest, largely because they develop on the right forward quadrant of the tropical cyclone and are steered by its counterclockwise circulation.

Despite their characteristically weaker intensities (they are generally categorized as EF0 to EF2), tornadoes generated by tropical cyclones are capable of significant destruction. Because they usually occur far from the center of the storm and move quickly, they can catch people by surprise. But more commonly, because the tornadoes are generally of the weakest type and because people have already evacuated or at least sheltered in sturdy buildings, fatalities related to tornadoes generated by tropical systems tend to be fewer than those from other tornadoes. Nevertheless, it is important to be protected from tornadoes that may touch down well before the direct effects of tropical cyclones are felt.

Hurricane Agnes's Tornadoes

Hurricane Agnes struck Florida with the fifth-deadliest tornado outbreak in Florida history in June 1972. As Agnes traveled north, it created the costliest natural disaster in U.S. history at the time; the damage from this storm totaled $3.5 billion (in 1972 dollars). Rainfall from Agnes created extreme flooding in Pennsylvania, New York, Maryland, Virginia, and Washington, D.C., that killed 122 people. Researchers Hagemeyer and Spratt (2002) examined the deadly tornadoes that Hurricane Agnes spawned in Florida. On June 18–19, 1972, the storm produced 28 tornadoes over the southern half of the peninsula, from just south of Daytona Beach to Key West. Two of the tornadoes produced F3 damage, nine were ranked as F2, eleven were ranked as F1, and the other six were ranked as F0. One tornado cut a path 100 yards wide through Okeechobee City, killing six people and destroying 50 mobile homes. Another tornado near La Belle in Hendry County killed a woman in a trailer. The tornadoes injured 140 people and destroyed 15 frame homes and over 200 mobile homes.

Hurricane Ivan's Tornadoes

Over 100 tornadoes occurred across several eastern U.S. states from September 15 to 18, 2004, in association with Hurricane Ivan. An F1 tornado touched down in downtown Panama City, claiming one life and damaging many structures. Several F2 tornadoes occurred on the same day in more rural areas of Florida. This event serves as a classic example of how numerous and distant from the eye of the storm tornadoes can occur.

Summer Tornadoes

One type of tornado development that is common during the summer occurs when air near the earth's surface is warm and this warm air rises rapidly into a cloud updraft into colder air aloft. This process is similar to the process that makes a hot-air balloon rise. It typically occurs at intersecting low-level boundaries, such as when a sea breeze merges with a thunderstorm outflow boundary, forcing the warm surface air upward as the parent cloud grows rapidly overhead. The intersecting boundaries create a low-level circulation that narrows and intensifies as the air rises to form a funnel. Sometimes these funnels protruding from a cloud base do not contact the ground or water; in such cases they are called funnel clouds. Some tornado funnels do stretch to the surface but might not be visible all the way to the ground or the water. The shape of the coastline often affects the likelihood of these tornadoes. Areas with peninsulas or bays can create a convergent surface wind flow that leads to stronger storms and a greater possibility for tornado development, while areas with a convex coastline can lead to divergent flows and weaker storms. Some summer tornadoes form over the warm waters as a waterspout. Figure 6.4 shows a waterspout over the St. Johns River near Jacksonville that developed during the summer of 2009 and moved onshore, toppling trees onto automobiles.

Figure 6.4. A waterspout over the St. Johns River near Jacksonville that developed during the summer of 2009 and moved onshore, toppling trees onto automobiles. Source: National Weather Service and NOAA (2009).

The Tornado and Waterspout of July 12, 1995

On July 12, 1995, a tornado developed rapidly over Pinellas County as outflow boundaries intersected, creating a low-level circulation and a strong rising motion that spun up the tornado. The tornado moved toward the south-southeast, over St. Petersburg, and left a path of damage that was estimated to be an F1 on the Fujita scale. It then moved across

Lake Maggiore and offshore over Tampa Bay, where it became an enormous waterspout (Collins et al. 2000).

Warm-Season Tornadoes in Southwest Florida

Collins, Paxton, and Williams (2009) discovered an interesting pattern that leads to warm-season tornadoes over southwest Florida. The intersection of sea breezes and outflow boundaries initiates a convective updraft that sometimes leads to a weak non-supercell tornado. Outflow boundaries are created by the downward rush of rain-cooled air from thunderstorms. In southwest Florida, warm-season tornadoes often develop as waterspouts and then move onshore. Since Florida's population density is greatest adjacent to the coast, as the tornado makes landfall, localized damage to homes often occurs. Depending on the wind direction, wind flow in the first several thousand feet above the ground pushes or restrains the circulations of the sea breezes that develop along the east and west coasts of the Florida Peninsula. The wind flow and the sea breezes eventually merge and initiate violent thunderstorms. The research team noticed an interesting pattern that signals tornado development. When east-to-west (easterly) wind flow occurs on the surface and north-to-south (northerly) winds are present in the upper atmosphere, the sea-breeze interactions along the irregularly shaped coastline create a broad area of low pressure as thunderstorms develop over southwest Florida. This area of low pressure precedes nearshore waterspout or tornado development by up to one hour. These types of indicators may increase the ability of meteorologists to predict tornadoes in this part of the state.

Cool-Season Tornadoes

The strongest tornadoes occur during the cool season when cold fronts and rapidly moving air masses create strong winds aloft and much colder air is present in the middle levels of the atmosphere. Unlike Florida's weaker and short-lived warm-season tornado-producing thunderstorms, storms that occur during the cool season have the ability to strengthen, rotate, and proliferate within a single storm system. Several deadly outbreaks have occurred in Florida in the recent past and are examined in the following sections. Grazulis (1993) defined the term "tornado outbreak" as a group or family of six or more tornadoes spawned by the same general weather system.

The March 1962 Tornado

On March 13, 1962, a savage tornado ripped through Milton, Florida (located about 30 miles northeast of Pensacola), claiming 17 lives and injuring 80 (Associated Press 1962). This was the highest death toll from a tornado in Florida to that date. The tornado devastated three residential blocks, destroying homes and leaving only stumps where trees had been. It picked up a home with three residents inside and took them for the ride of their life: it spun the house and gently dropped it 100 yards away on the foundation of another home that had been demolished by the twister, leaving the residents of the first home unharmed. Relief workers, members of the National Guard, and sailors from nearby Whiting Field poured into the area to help the victims.

The April 1966 Tornado Outbreak

Eleven people died and over 3,300 people were injured during the 1966 tornado outbreak that left a path of destruction across central Florida for over 100 miles (160 km). This occurred during the morning of April 4, 1966, beginning around 8:00 a.m. The strongest tornado, which had an estimated intensity of F4, created a path of death and destruction from Clearwater on the west coast of central Florida to Merritt Island, near the Kennedy Space Center, on the east coast. At times, it was 300 yards wide. In addition to the deaths and injuries, the storm destroyed more than 250 homes and businesses.

An article in the *St. Petersburg Times* (Davis and Millott 1966) reported that the tornado ripped through parts of the Forest Hills and Carrollwood areas and through the University of South Florida campus in Tampa. One citizen said that he saw the furniture literally sucked out of a home. East of Tampa, in Lakeland in Polk County, seven people lost their lives and a 55-foot radio tower was yanked from its concrete pilings and smashed to the ground. As the storm clouds continued moving eastward, citrus trees near Auburndale, Florida, were stripped bare and fruit was left on the ground. The second tornado moved ashore near the mouth of Tampa Bay and created destruction for over 100 miles to Cocoa Beach, where over 20 frame homes, a shopping center, and 150 mobile homes were destroyed and over 100 people were injured. This tornado outbreak is the fourth deadliest to be documented in Florida. All three of the deadliest tornado outbreaks occurred in winter to early spring where 42 people were killed in February 1998, 21 were killed in February 2007, and 17 were killed in March 1962 (NOAA National Centers for Environmental Information n.d.b).

The March 1993 Superstorm and Tornadoes

The March 1993 superstorm, known as the storm of the century, struck the Gulf Coast of Florida late on Friday, March 12, 1993, and continued to slam Florida and states to the north on the following day. The series of images in figures 6.5a–f show the evolution of the storm within one day, from a tattered collection of clouds over Texas to a huge storm menacing the eastern United States. Why was it called the storm of the century? To Florida residents, it was an unnamed and unprecedented March storm of an intensity that rivaled that of a hurricane; its wind gusts reached over 90 mph and it included tornadoes and a devastatingly deadly storm surge. But it was much larger in diameter than a hurricane and unlike a hurricane, it had a cold core rather than a warm core. People farther north called this storm the blizzard of the century; it was a blizzard like few had seen. It dropped temperatures, dumped snow, broke trees, and knocked out power over a wide swath from Georgia to Maine. The superstorm caused over $2 billion in property damage across portions of 22 eastern U.S. states. Much of the property damage occurred in Florida. Advance warnings saved lives. Amazingly, fewer than 100 deaths were directly attributable to the storm, and half of these occurred on vessels in seas estimated as high as 65 feet. Another 118 people perished from indirect causes; many of them died during the post-storm cleanup (NOAA and National Weather Service 1994).

Five days before the storm, computer models indicated the development of an intense low-pressure system over the Gulf of Mexico as it rocketed from coastal Texas to Florida then up the eastern seaboard. It was initially difficult for forecasters to believe that a weak low-pressure

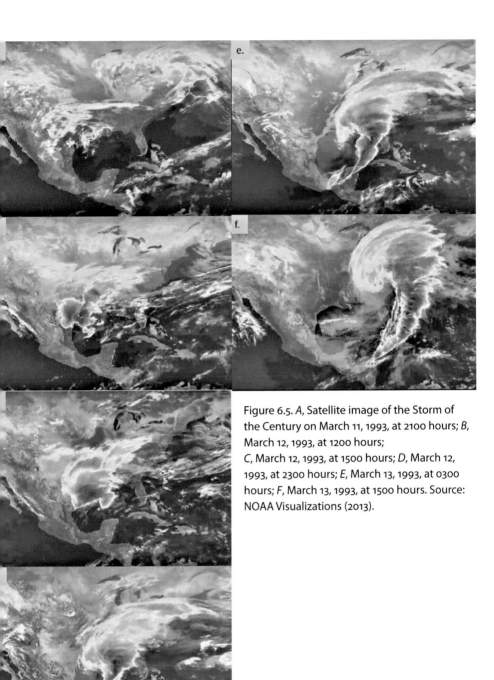

e.

f.

Figure 6.5. *A*, Satellite image of the Storm of
the Century on March 11, 1993, at 2100 hours; *B*,
March 12, 1993, at 1200 hours;
C, March 12, 1993, at 1500 hours; *D*, March 12,
1993, at 2300 hours; *E*, March 13, 1993, at 0300
hours; *F*, March 13, 1993, at 1500 hours. Source:
NOAA Visualizations (2013).

area could intensify so rapidly. Some forecasters described what they were seeing in the numerical models as a "meteorological bomb." As the week went on, the numerical forecast models continued to show the same unbelievably intense rapid development. Upstream, the upper-level flow produced a series of bends toward the equator that meteorologists had described earlier as troughs; these troughs were embedded in a larger trough. Disturbed weather usually occurs beneath the eastern, or right, side of the upper trough, where warmer, and usually moister, air begins to be pushed toward the closest pole and interacts with much colder air. Florida and the eastern third of the nation was situated on the stormy side of the main trough. In addition, the Arctic, polar, and subtropical jet streams were merging in the upper-level flow—a sure sign of a disturbance—and a deep flow of tropical moisture over the Gulf was coming north from the Caribbean Sea. These merging factors set the timer for the impending "meteorological bomb" that grew into a monster on March 12, 1993. The winds howled as the storm moved northward. According to NOAA and the National Weather Service (1994), the strongest recorded wind gusts were in Mount Washington, New Hampshire (144 mph); Franklin County, Florida (110 mph); Dry Tortugas, Florida (109 mph); and Flattop Mountain, North Carolina (101 mph).

A fast-moving squall line of thunderstorms produced 59,000 cloud-to-ground lightning flashes; the highest flash density was just south of Tampa as the line moved onshore along Florida's Gulf Coast. The superstorm created an unprecedented storm surge of up to 12 feet in Taylor County, located well north of Tampa Bay. The surge drowned 13 people.

Numerous tornadoes struck the state, including an F2 tornado near Chiefland in Levy County that caused three deaths. Other F2 tornadoes occurred along a 30-mile track in Lake County, causing one death there and another death near Ocala in Marion County. F1 tornadoes occurred near La Crosse in Alachua County, causing 1 death; near Crystal River in Citrus County; and in Jacksonville in Duval County. F0 tornadoes touched down near Treasure Island in Pinellas County, in New Port Richey in Pasco County, in Tampa in Hillsborough County, in Bartow in Polk County, and in Jacksonville.

Charlie Paxton was the forecaster on duty at the Tampa Bay National Weather Service during the day on Friday, March 12, 1993. He returned to the weather office that evening to issue warnings for the event. He recalls working the storm that night:

When I arrived at the office, satellite imagery showed the squall line racing east at 70 mph! Our team issued 26 severe thunderstorm warnings with lead times ranging from 30 minutes to over two hours! Standard warnings usually indicate wind gusts over 55 mph, but I upgraded wording in all of the warnings to indicate winds of over 90 mph. Of the 6 tornadoes in our area ... lead times were all over 20 minutes with the longest lead time of 48 minutes. Remember that we were using the old Weather Service Radar-57 Radar (57 as in 1957). We didn't have a local Doppler radar. We had a processor attached to the radar called RADAP and I had written software to make calculations on the severity of cells and that really helped.

We used an XT PC to send products through our main communication system called AFOS. We communicated with the Melbourne WSR-88D operator who helped identify tornadic circulations within range of their radar. We used the NAWAS line to communicate with the State Warning Point and county Emergency Operations Centers. We also received a number of reports from the local media. We had an 800 number available to the public and our phone didn't stop ringing. People were shocked at the intensity of the storm and provided us with many accounts of damage.

An official NOAA Service assessment offered these words about the performance of the Tampa Bay National Weather Service office: "Review of the verification statistics show clearly that excellent warning service was provided by WSO Tampa. Of the 26 severe weather warnings issued, 22 were verified. Of the 45 reported events, 43, or 95 percent occurred with either a Severe Thunderstorm or Tornado Warning in effect prior to the event. These results are exceptional for any NWS office" (NOAA and National Weather Service 1994).

Tornado Outbreak of February 1998 in Central Florida

Below is an account of the deadliest tornado outbreak in Florida in February 1998 from the National Weather Service in Melbourne.

During the late night and early morning hours of 22–23 February 1998 (Sunday–Monday), the most devastating tornado outbreak ever to occur in the state of Florida in terms of both loss of life and property damage, occurred from Kissimmee to Sanford to Daytona Beach. Forty-two people died as a result of the tornadoes and more than 260 others were injured. Over 3,000 structures were damaged, and more than 700 were completely destroyed. A total of seven confirmed tornadoes occurred that night. Four of the tornadoes were unusually long-lived and produced damage tracks of between 8 and 38 miles, resulting in the majority of damage and all fatalities. Uncommon for Florida tornadoes, the estimated wind speed for three of these twisters reached 200 mph which is on the high end of F3 intensity on the Fujita scale [and would be on the border between EF4 and EF5 intensity today]. (National Weather Service Forecast Office, Melbourne, Florida 1998)

The Groundhog Day Tornado of 2007

Ed Frederick (2009) chronicled survivors' stories from those who lived through the tornado outbreak in The Villages, Lady Lake, and Lake Mack on Groundhog Day 2007 in the book *Ten Seconds inside a Tornado*. Roger and Pat Rylott described their ordeal as they were awakened by a tornado warning alarm blaring on their weather radio.

Within six minutes upon entering the closet, we are in the midst of the unmistakable ground vibrations and sound of a tornado. We say "I love you," and struggle to hear every sound beyond the closed closet door. There is little time to wait. Windows breaking, a sand blaster on the other side of the closet door, groans, moans, the whine of winds slamming doors, and large items smashing into surfaces, such noise instills an imprint in our minds that will surely last a long time. Grass, glass, and palm boughs flush under the closet door and hit my leg as I hold the door shut; a substantial force pushes and pulls on the other side. There isn't time to dwell on getting hurt—or worse. Only ten to twelve seconds pass and . . . all is calm[,] too calm. As we stand up we realize that we are very lucky. (Frederick 2009, 293)

The tornadoes that touched down that day took 21 lives and injured another 76 people. In the early morning hours, meteorologists issued special marine warnings and then a tornado warning for a strong circulation seen on Doppler radar associated with a thunderstorm complex moving onshore on Florida's west coast and into Citrus County. The first tornado warning expired as the suspect circulation weakened with no reports of damage. Then, within minutes, the circulation gained strength and another tornado warning was issued. Minutes later, residents of The Villages community who were not wakened by the siren sound of the warning on their NOAA Weather Radios awoke to the horrendous sound of their homes being destroyed. The tornadoes demolished 200 homes and damaged over 1,100 in Sumter County. Those were the lucky victims—all they lost were their homes. The tornado continued moving eastward at 60 mph with over 160 mph winds, ravaging the Lady Lake area, killing eight people and destroying over 100 homes in Lake County before it lifted. A second tornado developed minutes later and obliterated the Lake Mack area, killing 13 people and destroying over 500 homes. A third tornado touched down just east of the second tornado before lifting along the east coast at New Smyrna Beach. The outbreak was the second deadliest on record for Florida. It caused $218 million in damages.

The Tornado's Cousins

Tornadoes have many cousins that also have a spinning vortex appearance, such as dust devils and waterspouts. The fast-swirling funnel shape efficiently equalizes imbalances in the atmosphere. These cousins occur in different weather scenarios, but most have the same fundamental reason for development: a heat imbalance between the warm surface of the earth and colder temperatures above that leads to rapidly rising air. When tornadoes form over water, they are termed waterspouts.

Waterspouts, like tornadoes, form when air is rising rapidly within a cumulus or cumulonimbus cloud. Funnels form more easily over the water because the frictional effects are less there than they are over land, where trees and buildings impede the more delicate circulations. Thus, most waterspouts are typically not as strong as tornadoes.

Sometimes strong wind in the lower atmosphere near the ground can create a spinning vortex. A gustnado is a short-lived whirlwind that lasts from seconds to a couple of minutes. It resembles a tornado but is caused by a downburst of air from a thunderstorm. The term "gustnado" is seldom used in Florida, but migrants from the midwestern states are likely to have heard it before. Gustnadoes can cause injuries and damage and are often simply considered tornadoes.

Florida meteorologists sometimes receive complaints on clear, sunny days of roofing shingles being ripped off and tossed about. What could be causing the damage? The answer is attributed to dust devils, which are generated on warm days when the ground surface is hot and the air aloft is cooler, creating thermodynamic instability. When these kinds of vortices occur in association with forest fires, they are called fire swirls. Fire swirls may cause a fire to intensify.

Not all tornado-like circulations spell trouble. A sometimes benign and often intriguing cousin of the tornado is the steam spout or steam devil, which forms on cold calm days over warmer bodies of water such as ponds, lakes, and rivers. This phenomenon is often accompanied by steam fog, which will be described in chapter 8. It is easy to see how indigenous peoples may have considered steam spouts to be visits from the gods.

7

Hurricanes

The word "hurricane" is one of the first things that comes to mind when people think about the weather in Florida. Florida's location makes it a target for hurricanes. Only storms of the most powerful category of tropical cyclone in the Atlantic and Eastern Pacific Oceans are called hurricanes. Tropical cyclone is a generic name for well-organized low-pressure systems or storms that form in or near the tropics. In the western Pacific Ocean, the most powerful tropical cyclones are called typhoons rather than hurricanes. Because the part of the Indian Ocean where tropical cyclones exist lies nearly entirely in the tropics, there they are simply called cyclones. The name "hurricane" is from the Spanish word *huracán*. The Spanish explorers derived the word from *hurakán* ("god of the storm") used in the Taino language of the Arawak people, who formerly inhabited the Greater Antilles and the Bahamas. It appears that the Arawak or other Carib Indians derived the word from the name of the Mayan god Hurakan, a creator god who blew his breath across the chaotic water. Regardless of what they are called around the globe, tropical cyclones are one of the deadliest types of storms on earth.

The Ingredients of a Tropical Cyclone

Compared to the mid-latitude cyclone, which derives its energy from contrasts in air temperature, tropical cyclones derive their energy from the ocean's heat and moisture as air rises and condenses and the latent heat of vaporization is released. With their characteristic general westward, then poleward movement, tropical cyclones are a natural mechanism for moving excess heat energy from the tropics toward the poles. The strength of tropical cyclones depends on the availability of three primary conditions: ocean waters over 80°F, moisture that extends through great heights in the atmosphere, and little or no wind shear. As summer approaches, the tropical and subtropical waters of the Caribbean Sea and the Gulf of Mexico warm the fastest. This is where tropical cyclones are likely to develop early in the season. As the season progresses, the waters of the Atlantic Ocean warm and the development area extends eastward to the coast of Africa. The warm water gives incipient storms heat energy. But if the warm water consists of only a shallow layer above a much colder layer, the storm's winds will churn up cooler water from below the surface and weaken the storm.

A moist atmosphere is needed because it provides latent heat to the storm. When energy from the sun is used to evaporate the warm water near the tropical cyclone, the evaporated water swirls up around the eye and cools and condenses back into a liquid. The same amount of energy that was needed to evaporate the water is then released back into the environment in the condensation process, providing additional fuel for the hurricane. Amazingly, so much water is involved in condensation around a hurricane that huge cloud decks form across hundreds of miles. In contrast, when a tropical cyclone moves into an area with drier air, the system weakens. During the summer, strong winds over the Sahara pick up dust and this drier air blows off the coast of northern Africa over the Atlantic Ocean. This layer of dry and dusty air, known as the Saharan Air Layer (SAL), generally has a negative impact on developing tropical systems. At times, the trade winds carry the Saharan dust over Florida, changing the typically vibrant blue sky to a milky color overcast with a tinge of brown.

The third condition needed for tropical cyclone development is the absence of strong vertical wind shear: wind speed and/or wind direction around the periphery of a developing storm does not change much with height. When the winds move at the same speed and in the same direction of a tropical system, the vertical development of a tropical

Figure 7.1. Satellite image of Tropical Storm Alberto in 2012, a weak storm. Source: NASA Worldview.

system continues unimpeded, leading to stronger tropical cyclones. When strong wind shear exists in the opposite direction of the cyclone's movement, the storm will lose its symmetric shape and weaken. In 2012, Tropical Storm Alberto had just a few anemic bands of low clouds swirling into the center and little vertical development after it weakened as it moved across the southeastern states and encountered wind shear over the Atlantic Ocean (see figure 7.1).

Where Tropical Cyclones Form

As the name implies, tropical cyclones form in the tropics, over the ocean areas, typically between 10° to 35°N latitude and 10° to 35°S latitude. Near the equator, the Coriolis force, an effect related to the rotation and curvature of the earth, is not strong enough to create the circulation of a cyclone. Map 7.1 shows the paths of tropical cyclones across the planet. In the Northern Hemisphere summer, the North Atlantic, the eastern and western North Pacific, and the Indian Oceans become

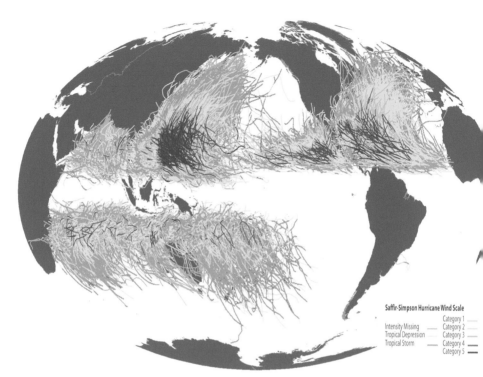

Map 7.1. Global areas where tropical cyclones develop and propagate. Source: NOAA Climate.gov (2010).

active tropical cyclone breeding grounds. In the summer in the Southern Hemisphere, only the western Pacific and Indian Oceans are active tropical cyclone regions.

Atlantic tropical cyclones that form near the coast of Africa and make the long track westward toward North America are called Cabo (Cape) Verde storms. These storms are often more powerful because they have more time to tap into the abundant energy of the warm, deep Atlantic Ocean and to develop a more organized and efficient circulation. However, non–Cabo Verde storms can also be dangerous because they leave less time for coastal dwellers to prepare.

Regardless of where they form, tropical cyclones are usually steered toward the west by the trade winds in the tropical latitudes as they develop. At some point, though, the storms typically move poleward, and generally as that happens, they begin to lose their tropical characteristics and become extratropical as they encounter cold fronts, drier air, and/or cooler waters. As this happens, the strongest winds, which are near the center of the storm in a tropical cyclone, begin to migrate and spread out away from the center. These extratropical storms can slam mid-latitude areas of the world, including Japan, Alaska, Newfoundland, the rest of Canada's Maritime provinces, and even the European continent.

Characteristics of Tropical Cyclones

Tropical cyclones are low-pressure weather systems with cyclonic in-flowing surface winds that form in or near the tropics. Unlike mid-latitude low-pressure systems, tropical cyclones have a warm core near the center of circulation, even in the upper levels. The wind field, which is strongest near the center of tropical cyclones, is another characteristic that sets them apart from mid-latitude cyclones. Intense lightning is rare in tropical cyclones because they are relatively warm in the upper levels and ice particles are an important part of lightning initiation. Anyone who has experienced a hurricane has probably noticed that lightning during tropical cyclones is indeed rare. As tropical cyclones intensify, a clear area, or eye, may appear near the center of the circulation (figure 7.2). The eye is an area of sinking air and light winds at the center of the

Figure 7.2. Tropical cyclone structure. Credit: Melissa Metzger.

Figure 7.3. Eyewall of Hurricane Katrina from within the eye. Source: NOAA 2005.

storm. It is surrounded by rapidly rising air that moves counterclockwise in the Northern Hemisphere. The area surrounding the eye is known as the eyewall and is typically the location of the strongest winds and the most intense severe weather. Figure 7.3 shows a view of the eyewall from within the eye of Hurricane Katrina in 2005. Near the top of a well-developed tropical cyclone, the wind flows away from the eye in all directions. This upper anticyclonic outflow creates a stronger mechanism for the increased upward flow of air from the lower part of the storm.

Table 7.1. The Saffir-Simpson scale

Hurricane Category	Wind Speed (mph)
1	74–95
2	96–110
3	111–129
4	130–156
5	≥157

Hurricane Categories

Most hurricanes evolve from tropical disturbances, which are discrete systems of organized convection (showers or thunderstorms) that originate in the tropics or subtropics, do not migrate along a frontal boundary, and maintain their identity for 24 hours or more. When a tropical disturbance gains a cyclonic circulation, the system is deemed a tropical depression. When a tropical depression gains strength to a peak sustained wind speed of 39 miles per hour, it reaches tropical storm status and is given a name. When it grows to a peak sustained wind speed of 74 mph, the storm is given hurricane status. Hurricanes are categorized by intensity using the Saffir-Simpson scale, which ranges from Category 1 to Category 5. A Category 5 hurricane is capable of completely destroying property (Table 7.1). The Saffir-Simpson scale categories correspond to the wind speed that begins to damage categories of sturdy structures. For example, as a hurricane becomes a Category 2, it is likely to damage roofs, windows, and doors. A Category 3 or higher hurricane is also known as a major or intense hurricane.

Names of Hurricanes

The naming of hurricanes has evolved over the years. In the late nineteenth and early twentieth centuries, some Caribbean hurricanes were named after saints. An Australian meteorologist named tropical cyclones after women in the nineteenth century. Until the 1950 hurricane season, Atlantic storms weren't given names. In 1950, U.S. forecasters began using the phonetic alphabet (Able, Baker, Charlie) to name storms. Three years later, they began using female names for storms. It wasn't until 1979 that male names were alternated with female names in the Atlantic hurricane basin. Male names were included at this point because the feminist movement didn't want something as destructive as a hurricane to be associated only with women. Because storms impact people in countries of the Caribbean and Central America, where English is not the primary language, French and Spanish names were included beginning in 2003. The names are reused every six years unless a storm is so notable that the name is retired and a new name is chosen by a committee.

Hurricane Season

During some years, Florida becomes the Hurricane State. From 1851 to the present, the entire period when official hurricane records have been kept, Florida has experienced more tropical cyclone activity than any other state. Florida experiences twice as many land-falling hurricanes as Texas, the next most prevalent state. Of course, Florida also has more coastline than Texas.

Although some storms can form early and some form late in the season when the ocean waters are still warm, the official Atlantic hurricane season begins on June 1 and ends on November 30. The storms in June tend to form near Florida in the Gulf of Mexico or the Caribbean Sea, where the waters warm faster than other areas of the Atlantic basin (see map 7.2). Although storms that form close to home leave little time for preparation and evacuation, June storms are typically weak. As the season continues and the Atlantic Ocean waters warm to over 80°F, the storms form farther away from Florida. August and September are the peak months for hurricane activity in the Atlantic basin. Almost 80 percent of the tropical cyclones that have hit Florida have occurred in the period August through October. During the peak of the season, disturbances move off the west coast of the African continent over the warm waters of the Atlantic around 20°N and grow into tropical depressions. One-third of the storms that make landfall in Florida occur during September; the peak month for tropical cyclone activity. About 20 percent of Florida's tropical cyclones strike in August and 25 percent strike in October. By October, the sun is lower and cooler air filters

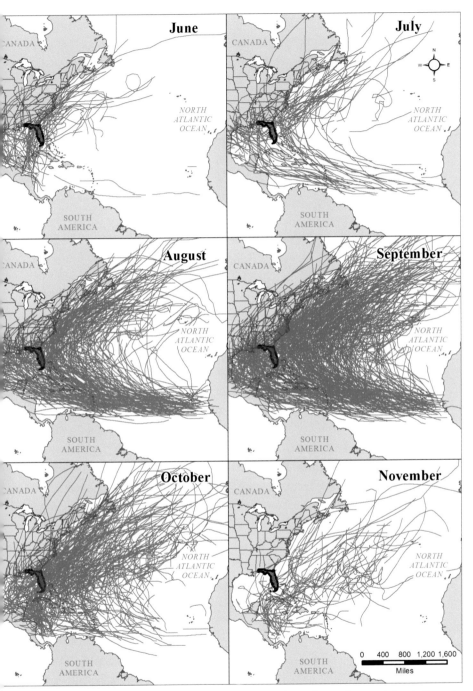

Map 7.2. Seasonal tropical cyclone development over the Atlantic basin. Source: Atlantic Oceanic and Meteorological Laboratory (2016).

southward behind cold fronts. Notice that October storms are likely to form near Florida and are often stronger than the June storms that develop in nearby locations. Only 4 percent of tropical cyclones in Florida have struck in November and December. Those storms are typically weak because the number of daylight hours is decreasing during these months and the waters surrounding the state are cooling. The remaining approximately 16 percent of the storms that impact Florida occur from January through July.

Hurricane Hazards

The tropical cyclones that have impacted Florida have ranged from very weak storm systems to the most powerful. Hurricane Andrew will be remembered forever for its obliteration of Homestead, Florida, and the southern parts of Dade County. But sometimes the weaker systems have also made the history books because of their impact. For instance, in September 1950, the weak but slow-moving Hurricane Easy set the 24-hour record for rainfall in Florida. Easy stalled over the Gulf of Mexico, picked up enormous quantities of water, and dumped over three feet (38 inches) of flooding rain in a 24-hour period over Yankeetown, Florida.

For coastal residents, the most formidable danger is the battering surge of rising turbulent water as powerful storms move onshore. Figure 7.4 shows the height that storm surge water could reach in a St. Petersburg neighborhood well away from Tampa Bay. For inland dwellers, the most damaging effect of a tropical cyclone is drenching rainfall that creates flash flooding capable of washing away homes and automobiles. In both coastal and inland areas, high winds can cause extreme damage as they toss automobiles around, break uncovered windows, and remove roofing. The storm surge, rain, and winds have the potential to completely destroy property and cause considerable loss of life. In addition, tropical cyclones are often accompanied by tornadoes.

The Most Powerful Side of the Storm

The movement of the storm and the counterclockwise rotation of the flow around the eye near the earth's surface make winds on the right side of the storm stronger and much more dangerous. The right side of

Figure 7.4. Sign indicating potential storm surge in St. Petersburg. Credit: Charles H. Paxton.

the storm is also where the wind will push storm surge from the ocean or gulf waters onshore. Storm surge is less of a factor where the winds will flow from land to sea, which means that the water and waves will be pushed offshore. Also, drier air may be present on the storm's left side at landfall and friction from the rougher terrain may slow the winds somewhat. When Hurricane Opal made landfall on October 4, 1995, on Santa Rosa Island just east of Pensacola, the Florida Panhandle beaches nearest the eyewall had storm surges of up to 15 feet, while the Mobile Bay area had storm surges of only around 4 feet (figure 7.5).

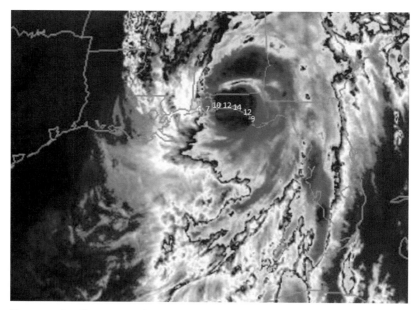

Figure 7.5. Satellite image of Hurricane Opal in 1995 and estimated storm surge heights in feet. Source: National Weather Service and NOAA (1995).

Frequency of Hurricanes

Because of the shape, orientation, and extent of coastline, Florida has a long history of tropical cyclone landfalls. It has been noted that a hurricane strikes the state every two years on average. However, after Wilma in 2005, Florida did not have another hurricane until the 2016 season, when Hurricane Hermine struck. The next-longest inactive period was from 1979 to 1985. That streak was ended by Hurricane Elena, which spun over the gulf waters for days then smacked the Panhandle with 92 mph winds and 10 feet of storm surge. Another break in hurricane landfalls began in 1987 and abruptly ended in 1992, when Hurricane Andrew mowed through south Florida. People often think of Florida as the hurricane state.

Because most hurricanes that form in the North Atlantic Ocean track westward to northwestward, the southeast region of Florida is more prone to hurricanes than the western peninsular region. However, the southwest coastline can get hit from hurricanes that form in the Gulf of Mexico and hurricanes that form in the North Atlantic Ocean or the

Caribbean Sea and head into the gulf. When storms form in or enter the Gulf of Mexico and begin moving northward, the Florida Panhandle is often in peril. Keim, Muller, and Stone (2007) noted that southern peninsular Florida, where tropical storm– or hurricane-force winds occur on average once every three years, is as susceptible to tropical cyclones as anywhere else on the Atlantic-Gulf Coast of the United States, except perhaps the Outer Banks of North Carolina. The researchers note that in contrast, northeastern Florida and Georgia and the area of western Florida around Cedar Key have been struck less frequently by intense hurricanes during the period when instruments have been recording hurricanes than other stretches of the Atlantic and Gulf coasts south of New York: a Category 3 or higher hurricane has occurred in these parts of Florida on average once per 100 years or more.

Hurricanes Frances and Jeanne struck nearly the same stretch of the southeast coast of Florida in 2004 in what some considered to be a fluke, since hurricanes rarely strike the same place twice in a year (map 7.3). Interestingly, though, over 100 years earlier, in 1871, two similar storms made landfall in nearly the same place. On August 17, 1871, a major hurricane made landfall near Palm Beach with winds around 120 mph causing several shipwrecks on the Atlantic coastline and destroying roofs on buildings in Jacksonville. Just over one week later, on August 25, 1871, the fourth tropical storm and second hurricane of the season hit near Palm Beach, causing at least one shipwreck along the coastline (Atlantic Oceanic and Meteorological Laboratory 2016).

Short-Range Hurricane Forecasting

Before the advent of widespread communication capabilities, hurricanes struck by surprise, often killing thousands. Imagine trying to analyze a hurricane over the ocean on a surface map around the turn of the twentieth century before aircraft and satellite reconnaissance. Air Force aviators flew the first missions into tropical cyclones in the 1940s. The Hurricane Hunters of the Air Force Reserve's 53rd Weather Reconnaissance Squadron based at Keesler Air Force Base in Biloxi, Mississippi, began flying into tropical storms and hurricanes in 1944. Naval aviators also flew missions into tropical cyclones from 1945 to 1975. The NOAA

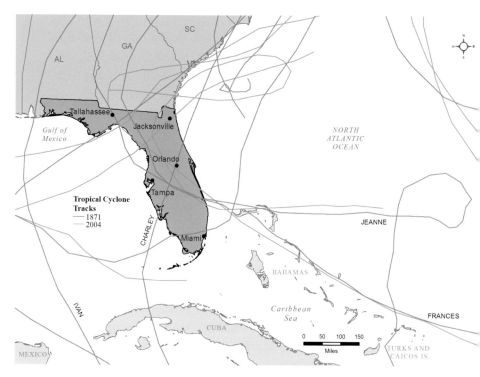

Map 7.3. Repeat hurricane landfalls in 1871 and 2004. Hurricanes Frances and Jeanne made landfall near Palm Beach in 2004, but over 100 years earlier, in 1871, two other hurricanes with similar tracks struck the same area. Source: Atlantic Oceanic and Meteorological Laboratory (2016).

also has a branch of the Hurricane Hunters. As Collins and Flaherty (2014) have noted, since January 1993, NOAA Hurricane Hunter aircraft have been operated and maintained by NOAA's Aircraft Operations Center in the Tampa Bay area. Currently operating under NOAA's Office of Marine and Aviation Operations, NOAA's Hurricane Hunters and its Aircraft Operations Center were born out of its Research Flight Facility, which operated out of the Miami International Airport from 1961 until Hurricane Andrew struck in 1992. The Hurricane Hunters perform surveillance, research, and reconnaissance with highly instrumented Orion P3 and Gulfstream G4 aircraft (figure 7.6).

The first satellite images showed the weather from space in 1960 but did not become available on a regular basis until around 1970 with the

Figure 7.6. G4 and P3 aircraft of the NOAA's Hurricane Hunters. Source: Department of Commerce, NOAA, and Office of Marine and Aviation Operations (2017).

first geostationary satellite sending images in 1975. As technology increased, the forecasts improved. As computer technology has advanced, hurricane modeling capabilities have become more detailed, accurate, and useful. Computer models provide better estimates of tropical cyclone tracks, allowing meteorologists to give more warning time than ever before. Television meteorologists often show "spaghetti" models that forecast the track of a particular storm using data pulled from a number of different numerical and statistical models. When the spaghetti is closely bunched, we have a better consensus and likely a more reliable forecast. This scenario is more likely when the storm is in an area of consistent steering winds. When the spaghetti is aligned in different directions, the forecast track is less reliable. This is more likely when storms are in an area of weak steering winds. Now it is possible to run the same model many times with slightly different initial conditions to obtain an even greater ensemble of results. When all the different ensemble model runs show similar tracks, meteorologists have a better consensus and can rely on the collective model output more. When the computer models provide a poor consensus, meteorologists may seek a particular model run that has produced the best verifiable short-term results.

Long-Range Hurricane Forecasting

Although seasonal forecasts of hurricanes are available and useful, it is important for Florida residents and visitors to bear in mind that they should plan for an active hurricane year every year. As an example, although 1992 was forecasted to be an inactive year, Hurricane Andrew, a Category 5 storm, struck south Florida on August 24 of that year, causing 26 direct deaths and a reported $25 billion (in 1992 dollars) in damage. Even in an inactive year, all it takes is one strong storm.

Both 2004 and 2005 were forecasted to be above average for Atlantic hurricane activity. In 2004, Florida was impacted by several landfalling tropical cyclones (Charley, Ivan, Frances, and Jeanne). Then in 2005, Florida was impacted by Dennis, Katrina, and Wilma. This caused people to wonder if we were entering a new era of increased hurricane frequency.

The Atlantic Multidecadal Oscillation

It is no coincidence that many historic hurricanes occurred in the 1920s and 1930s. Research has supported the notion that the Atlantic Ocean basin (including the Gulf of Mexico and the Caribbean Sea) undergoes periods of 20–40 years of abnormally warm conditions, followed by several decades of colder conditions. This is known as the Atlantic Multidecadal Oscillation (AMO). The warm phase of the AMO is associated with warmer than normal water temperatures in the Atlantic Ocean and an uptick in hurricane activity, as occurred in the 1920s through around 1960, when so many memorable storms happened. The 1960s through the mid-1990s, by contrast, was the cold phase of the AMO, and fewer storms hit Florida and elsewhere in the Atlantic basin in these decades.

Beginning in the mid-1990s, the warm phase of the AMO returned. Hurricane activity was relatively frequent in the North Atlantic from 1995 onward. The 1995 hurricane season was the busiest in the historical record until the 2005 hurricane season. In 2005, so many storms developed that all the alphabetical names were used and the other storms that developed were given Greek names. But a long, quiet period resumed after 2005, even though the AMO appeared to be in its warm phase. The absence of major hurricanes that hit the United States after 2005 appears to have been a streak of luck. According to a Weather Channel blog in 2015, hurricane expert Phil Klotzbach calculated the chance that 26 consecutive Atlantic major hurricanes would not make landfall in the United States as 1 in 7,400 (Dolce 2015). Although the Atlantic hurricane seasons have not all been particularly quiet since 2005 (for example, the 2010 season was an active year with 19 storms), none made landfall at hurricane strength in the United States until 2016. By way of explanation, Klotzbach and his colleagues suggested that the AMO flipped to the negative phase in 2012; other research has suggested that it may have been in the process of flipping during that time.

Social Issues

Even though the ability to forecast whether the AMO is switching cycles from warm to cold or vice versa would be extremely valuable, factors such as the frequency and intensity of storms represent only the physical dimensions of the event. The real hazard facing Florida and other weather-threatened locations is the complex web of social, economic, historical, and political forces that determine an area's vulnerability and, ultimately, the extent of its losses and its ability to recover. Research must focus on integrating the physical and human dimensions of hurricane impacts if we are to comprehend the dangers of extreme weather at a time when exposure and sensitivity to atmospheric hazards are increasing around the world.

Such understanding constitutes a major challenge for Florida. In 2015, the federal census estimated that the state had a population of 20 million people, a fivefold increase of 17 million from the 1950 population of three million, averaging over a quarter-million people arriving per year (U.S. Census Bureau 2015). This rapid population growth places more people at risk, many of whom may never have experienced severe weather, particularly those who have not experienced a direct hit from a landfalling hurricane in Florida. The vulnerability of the state's population is compounded by its demographic traits: it has the highest proportion of elderly residents, 11.6 percent of its people live in mobile homes, and many individuals are disadvantaged (U.S. Census Bureau 2015; U.S. Census Bureau 2011). Furthermore, the lack of experience with hurricanes, even for people who have been in Florida for decades, exacerbates the problem as residents are lulled into a false sense of security and quickly forget the impacts of hurricanes and the preparations that are necessary. However, modern hurricane-tracking technology and telecommunications have kept the number of deaths down, even in a state that is more densely populated than ever before.

Significant Florida Hurricanes

Many of the costliest mainland U.S. tropical cyclones have struck Florida. These include hurricanes Katrina, Andrew, Wilma, Ivan, Charley, Frances, Jeanne, Opal, Georges, Dennis, and Agnes. Map 7.4a shows major hurricanes (Saffir-Simpson Categories 3–5) that have hit Florida. Major hurricanes strike the United States every couple of years. In some years, more than one major hurricane has struck Florida, as happened in 1950, 2004, and 2005. In the twentieth century, the average time between major hurricanes that made landfall in Florida was four to five years. Table 7.2 shows the tropical cyclones that caused more than 25 deaths (map 7.4b). It is clear that more of the deadliest storms over the

Table 7.2. Storms that have struck Florida and caused 25 or more fatalities

Rank	Year	Storm Name	Category	Deaths
1	1928	Okeechobee	4	2,500
2	1935	Labor Day, Florida Keys	5	408
3	1926	Miami	4	372
4	1919	Florida Keys	4	287
5	1906	Southeast Florida	3	164
6	1896	Cedar Key/North Florida	3	130
7	1972	Agnes	1	122
8	1965	Betsy	3	75
9	1947	South Florida	4	51
10	1960	Donna	4	50
11	1910	Southwest Florida	3	30
12	1994	Alberto	Tropical storm	30
13	1992	Andrew	5	26

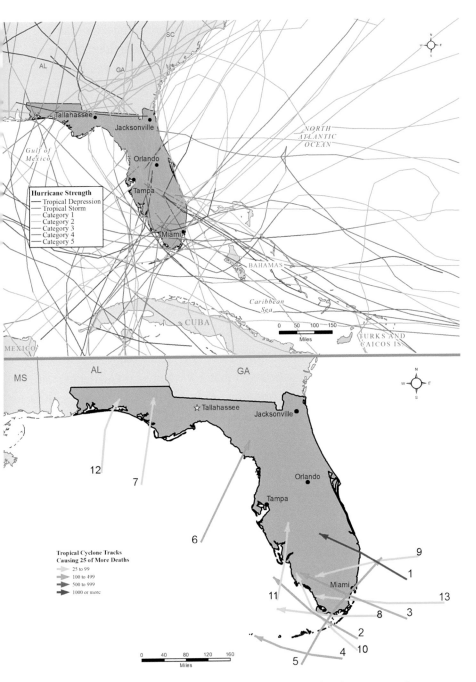

Map 7.4. *A*, Major hurricanes (Categories 3–5) that impacted Florida, 1851–2015; *B*,
Tropical cyclones that caused more than 25 deaths in the United States, 1851–2015.
Source: Adapted from Blake et al. 2007.

peninsula come from the east and that the deadliest storms over the Panhandle come from the south. The sections below highlight some of the most notorious Florida hurricanes in terms of fatalities and damage.

The Hurricane of 1848

A 15-foot storm surge in Tampa Bay reshaped much of the area when a major hurricane struck in late September 1848. Damage and loss of life were minimal because only a few people, mostly military, lived in the Tampa Bay region at the time. The surge had a huge impact on the fragile sandy barrier islands. The high surge, rough surf, and strong currents cut several gaps, or passes, through barrier islands, including Stump Pass at Englewood, Casey's Pass at Venice, New Pass at Palm Island (which became Longboat and Lido Keys), and John's Pass between Madeira Beach and Treasure Island (World Heritage Encyclopedia 2016).

The Dry Tortugas Hurricane (also known as the Florida Keys Hurricane) of 1919

This was one of the few historically severe hurricanes that was not a Cabo Verde storm. Instead, it is believed that it formed near Guadeloupe, in the Lesser Antilles. Its most important early impacts were in the Greater Antilles—Cuba, Jamaica, Hispaniola (consisting of the Dominican Republic and Haiti) and Puerto Rico—before it clipped the Florida Keys on September 10 as a Category 4 storm. As the storm progressed into the Gulf of Mexico, it attained one of its distinctive features—slow forward movement—before finally making another landfall in southern Texas on the evening of September 14. Because of this slow speed and wide circulation, this hurricane had tremendous storm surge impacts across most of the U.S. Gulf Coast, even in areas far from where it tracked. The storm caused approximately 800 fatalities from the Bahamas to Texas but less than 100 deaths in a sparsely populated Florida.

The Hurricane of 1921

The residents of the Tampa Bay area have enjoyed good fortune in dodging hurricanes for nearly 100 years. The last hurricane to impact the Tampa Bay area directly was an unnamed storm in October 1921. The estimated Category 3 storm made landfall just north of Tampa Bay, battered the city with winds exceeding 100 miles per hour, and created a 10- to 12-foot storm surge in Tampa Bay. The storm caused up to ten million dollars of damage and was responsible for six deaths (Ballingrud 2002). If a twin to this storm or a storm like Katrina were to hit the area today, public and emergency officials would need to act quickly to execute sound, detailed emergency plans to help evacuate vulnerable populations, as roughly a third of the current Tampa Bay population of 4 million lives at elevations of 10 feet or less.

The Okeechobee Hurricane of 1928

The Galveston (Texas) Hurricane of 1900 killed over 6,000 people, more than any other hurricane that made landfall in the United States to date. The second-place finishers in the record books are all too often forgotten. For instance, few people remember the second-deadliest hurricane to strike on U.S. soil, but this hurricane, the Okeechobee Hurricane of 1928, has secured its place in Florida history. Many of the characteristics of this great hurricane are recorded in the Hurricane Research Division's HURDAT database and in Neely's (2014) recent book on the topic. Before it made landfall in Florida, the Okeechobee hurricane was what is believed to be the only Category 5 hurricane to strike Puerto Rico since the age of European exploration. The Okeechobee hurricane also caused death and destruction in the Lesser Antilles, especially Guadeloupe. This fearsome hurricane was blamed for over 4,000 deaths, of which over 2,500 were Floridians, most of whom were located near its landfall area of West Palm Beach to Lake Okeechobee on September 17, 1928. The storm earned its "Okeechobee" moniker because its winds pushed so much of Lake Okeechobee's water into the adjacent lowlands, drowning many people who lived on the north and south sides of the lake. The track of this ferocious hurricane then turned northward, taking it lengthwise up the Florida Peninsula, which only added to its millions of dollars of destruction in the state. The storm's damage estimate is particularly notable because it occurred at a time when little economic development existed (Neely 2014).

The Labor Day Hurricane of 1935

The Hurricane Research Division of the National Oceanic and Atmospheric Administration (2014) and other scholarly sources also include information about the deadly Category 5 Labor Day hurricane that struck the Upper Keys on Monday, September 2, 1935, and killed over 400 people. At landfall, this storm may have had the lowest central pressure (892 mb) in U.S. hurricane history, along with 160 knot winds. One small silver lining is that the incredibly compact nature of the storm limited hurricane-force winds to perhaps within 25 miles of its eye at landfall, but the localized storm surge of up to 20 feet was devastating (McDonald 1935). This storm was comparable in strength and size to 2004's Hurricane Charley.

After striking the Upper Keys and moving across southern Florida into the Gulf of Mexico, the storm continued moving north but weakened just off Florida's west coast before making landfall again near Cedar Key on September 4 (map 7.5). This was the first of three Category 5 hurricanes to hit the United States in the twentieth century. A memorial plaque at a monument (figure 7.7) on the island of Islamorada describes the dreadful event:

The Florida Keys Memorial, known locally as the "Hurricane Monument," was built to honor hundreds of American veterans and local citizens who perished in the "Great Hurricane" on Labor Day, September 2, 1935. Islamorada sustained winds of 200 miles per hour (322 kph) and a barometer reading of 26.36 inches (66.95 cm) for many hours on that fateful holiday; most local buildings and the Florida East Coast Railway were destroyed by what remains the most savage hurricane on record. Hundreds of WWI veterans who had been camped in the Matecumbe area while working on the construction of US Highway One for the Works Progress Administration (WPA) were killed. In 1937, the cremated remains of approximately 300 people were placed within the tiled crypt in front of the monument. The monument is composed of native keystone, and its striking frieze depicts coconut palm trees bending before the force of hurricane winds while the waters from an angry sea lap at the bottom of their trunks. Monument construction was funded

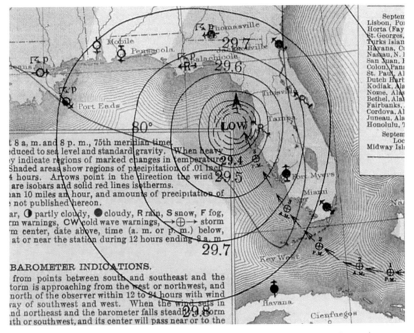

Map 7.5. The Labor Day hurricane, September 4, 1935. Source: NOAA Central Library Daily Weather Maps.

Figure 7.7. The Florida Keys Memorial, built to honor those who perished in the Labor Day Hurricane of September 2, 1935. Credit: Wikimedia Commons.

by the WPA and regional veterans' associations. Over the years, the Hurricane Monument has been cared for by local veterans, hurricane survivors, and descendants of the victims.

Hurricane Andrew (1992)

Hurricanes can occur even during cold phases of the AMO periods, and several hurricanes struck Florida in the 1960s through 1995. The most notable storm of this "quiet" era was Hurricane Andrew, a Cabo Verde storm that landed in southern Florida on Elliott Key and then at Homestead on August 24 as a Category 5. Hurricane Andrew became the costliest natural disaster in Florida history, with over $25 billion in damages (in 1992 dollars) (Landsea et al. 2004). At the time, Andrew left the most damage of any hurricane in U.S. history, though that dubious distinction has since been surpassed by Katrina (2005), Ike (2008), and Sandy (2012). Andrew's death toll of 44 Floridians was also substantial, but thanks to excellent weather forecasts, communication, and political leadership, the death toll was smaller than what might have been expected in such a catastrophic storm that made landfall at a place and in a time when few had any memory of hurricane preparation.

Hurricane Opal (1995)

The 1995 Atlantic tropical cyclone season will be remembered as one of the busiest since technology has existed to monitor hurricanes. Opal, one of the most memorable of the 1995 storms, caused more damage in Florida than anywhere else. Opal was unusual because it formed as a Cabo Verde storm but remained poorly organized until it reached just north of the Yucatán Peninsula, where it was named a tropical storm on September 30. Like most storms, Opal weakened substantially as it crossed land—in this case, the Yucatán—from east to west. Opal then rapidly strengthened to Category 4 status over the Bay of Campeche before ricocheting northeastward across the gulf and slamming into the Florida Panhandle near Santa Rosa Island on October 4. The storm killed 50 people in Central America and 13 in the United States. Only two deaths were in Florida, caused by an F2 tornado that Opal spawned (Centers for Disease Control 1996). The state bore a very heavy share of Opal's damage, which included major damage to Highway 98, which parallels the coast along Fort Walton Beach, Destin, and other resort areas. Opal served as an important reminder of the problems that accompany the aggressive and audacious development of coastal resort areas that took place from the 1980s through the mid-1990s. This development occurred during the cold-AMO period, when the absence of major tropical cyclone landfalls emboldened investors and developers to challenge nature. Sadly, humans learn too slowly that these are battles they cannot win (Webb et al. 1997; Centers for Disease Control 1996).

The 2004 Hurricanes: Charley, Frances, Ivan, and Jeanne

The decade-long period of relative quiet for Florida ended with a vengeance in 2004, when four historically powerful storms made landfall in or very near the state and another weaker tropical storm (Bonnie) struck the Panhandle. According to the National Weather Service, Ivan was the sixth-costliest hurricane to reach landfall in the history of the United States, Charley was the eighth-costliest, and Frances was the tenth-costliest. All of these storms affected Florida directly during this one season. It is notable that the first-, fourth-, and fifth-costliest hurricanes were also felt in Florida (Katrina and Wilma in 2005 and Andrew in 1992). The 2004 and 2005 seasons were amazingly costly to Florida in terms of loss of life and property.

Charley, a Cabo Verde storm of historic proportions, was the first major storm in Florida in 2004. It was known for its small geographic extent, for multiple landfalls across the Caribbean and the southeastern United States, and for setting a record (with Tropical Storm Bonnie) on August 13 (a Friday, no less) as the only pair of named storms to hit a single state within a 24-hour period. After Charley marched across the Dry Tortugas, an unseasonably early atmospheric trough dipped southward across the gulf, steering Charley northeastward. Charley intensified rapidly as it moved, before striking Cayo Costa along the southwest Florida coast with Category 4 winds. Often, the intensification of a tropical cyclone is accompanied by a shrinking eye and tighter circulation around the storm. Charley followed this pattern, which brought both good news and bad news to coastal residents. The good news was that the shrinking eye and eyewall meant that fewer coastal residents were subjected to the full brunt of the hurricane. The bad news was that the shrinking eye meant falling surface pressures and increasing wind strengths for those who were directly in its path. Charley continued inland over Florida, trekking over the heavily populated areas of Orlando, Kissimmee, and New Smyrna Beach while still at hurricane strength. Charley reemerged over the Atlantic Ocean and made additional landfalls in South Carolina, then merged with a frontal system as it moved over southern New England.

Frances, another Cabo Verde storm that is often forgotten because it was only a Category 2 at landfall near Hutchinson Island along the southeast coast, also caused significant impacts. Unlike Charley, Frances was a large, slow-moving storm that caused widespread damage but had less intense impacts. After wreaking destruction in the Bahamas as a Category 4 storm, Frances weakened before its Florida landfall and reemerged in the Gulf of Mexico near Tampa before turning back to the northeast and striking land again near St. Marks.

Unlike Charley and Frances, which struck on the Florida Peninsula, Ivan made landfall west of Pensacola, on the Alabama coast. But Ivan's large extent—its eye alone extended for more than 40 miles—caused its strongest impacts to occur in northwestern Florida, on the right side of the storm's landfall.

Although Ivan was a Category 5 storm as it entered the Gulf of Mexico, it weakened substantially to Category 3 strength just before its landfall on September 16 in Gulf Shores, Alabama. But by September 20, Ivan had made a huge clockwise loop, reemerging in the Atlantic before crossing the southern Florida Peninsula and making landfall near Miami. After that second landfall, it crossed the peninsula, reemerging in the Gulf of Mexico, finally making its third landfall in the mainland United States near the Louisiana-Texas border as a tropical depression. One of the most distinctive features of Ivan was its strength at such low latitudes. Ivan was a Category 3 storm while its eye was only 10.2 degrees of latitude north of the equator. Typically, tropical cyclones do not form within about 5 to 10 degrees of the equator because the Coriolis force is too weak at equatorial latitudes to allow tropical low-pressure areas to develop the characteristic spiraling component of their winds. Instead, near the equator, winds tend to converge linearly along the intertropical convergence zone. Ivan is among the ten Atlantic hurricanes with the lowest pressure in recorded history; its minimum pressure was only 910 mb. To area residents, the most memorable damage was to the Escambia Bay Bridge on Interstate 10, where huge segments of roadway were moved off the foundation by the storm surge. The height of storm surge varied widely along Ivan's path because of the various types of coastline configuration and water depths it encountered, but areas near the Bay Bridge experienced the worst, at over 13 feet (Stewart 2014).

Jeanne, which was not a Cabo Verde storm, followed Ivan into Florida by less than a week. Jeanne's Florida landfall occurred on September 26 near Hutchinson Island, almost the same location as Frances's landfall, but it quickly dissipated after moving inland and trekking north-northwestward up the western side of the Florida Peninsula. Like Ivan, Jeanne emerged in the Atlantic after crossing Georgia, the Carolinas, Virginia, and Maryland. However, unlike Ivan, Jeanne had already made its loop before reaching Florida and did not reappear for a second landfall. Jeanne's impacts were magnified by its timing—the rain from Jeanne had fallen on ground that was already waterlogged by Ivan, exacerbating flooding and crop damage. In terrain like Florida's, large trees are easily destabilized by such soggy conditions, particularly because the roots tend to be shallow because of the naturally high water table.

Katrina and Wilma (2005)

While Katrina will always be remembered for its devastation in Louisiana, it also had impacts in Florida. In southern Florida, Katrina caused 14 deaths, three of which were due to falling trees and another three due to drowning. On August 25, 2005, Katrina dropped over 16 inches of rain and created widespread damage and power outages. It also caused a tornado on Marathon Key that day. The pressure had dropped to 985 mb and the intensity had increased abruptly just before landfall, leaving many residents unprepared even though the National Weather Service had anticipated this strengthening. A huge sigh of relief ended the rapid preparations from Pensacola to Panama City. But as Florida's woes ended, the trauma was just beginning for people to the west of Florida. Katrina triggered some eerie flashbacks to Andrew in the minds of long-term residents of south Florida. Coincidentally, the storms shared similar tracks and had impacts in heavily populated areas, first in Florida and then in Louisiana. Andrew and Katrina even made landfall at almost the same calendar day in both places. The federal emergency response measures to both storms were unprecedented, led by President George H. W. Bush in the former and President George W. Bush in the latter. The aftermath of both storms showed remarkable acts that ran the gamut of human capabilities, including both courage and kindness, and greed and crime.

One of the storms with even lower central pressure than Katrina was Wilma during late October 2005, which holds the record low pressure in the Atlantic Ocean Basin. Wilma dropped almost 100 mb in one day to 882 mb. This broke the previous record of 888 mb, held by Gilbert of 1988. Wilma was one of the few intense hurricanes that approached peninsular Florida from the west, after having clipped Mexico's Yucatán Peninsula. This northeastward turn occurred because Wilma's late-season lifespan made it more vulnerable than most storms to being steered by the eastern side of an upper-atmospheric trough. Like Charley in the year before, Wilma remained near that cold-warm boundary that steered the storm into south Florida making landfall near Cape Romano, Florida, on Monday, October 24, 2005 (Hurricane Research Division 2004).

Hurricane Matthew (2016)

Hurricane Matthew became a Category 5 storm, ending an almost ten-year-long "drought" of these intense storms in the North Atlantic; the last Category 5 in this ocean was Hurricane Felix in 2007. Matthew also became a Category 5 at the southernmost location in the North Atlantic on record. Furthermore, Hurricane Matthew was the first hurricane to impact the east coast of Florida since Hurricane Wilma in 2005. According to the National Hurricane Center, Hurricane Matthew killed over 500 people in Haiti where it made landfall as a Category 4; it then brushed the U.S. east coast as a major hurricane (Category 3), moving along most of Florida, until it weakened to a Category 2 about 50 nautical miles east-northeast of Jacksonville Beach, Florida (NHC, 2017). On October 7, hurricane force winds were felt in Florida. Many people had evacuated the east coast of Florida (many headed toward the west coast), but Collins et al. (2017) noted that many along the coast who were under a mandatory evacuation order did not evacuate. While many locations on the east coast of Florida had damage, further damage was prevented because the westernmost edge of the eye wall was far enough out to sea when Matthew brushed the state. Matthew was the deadliest Atlantic hurricane since Hurricane Stan in 2005. In Florida, only two people directly lost their lives, due to falling trees, yet there were nine indirect deaths, including two people who succumbed to carbon monoxide poisoning caused by operating gas-powered electrical generators in their homes (NHC, 2017). Over one million people lost power in Florida, with more people losing power as the storm traveled up to Georgia and the Carolinas. Matthew was the costliest Atlantic hurricane since Hurricane Sandy in 2012.

Eyewall Replacement

Opal, Frances, Katrina, and Wilma each underwent eyewall replacement, a process that happens in some intense hurricanes but is only partially understood. Eyewall replacement involves a momentary weakening of the storm as the area surrounding the eye—the eyewall—weakens as an outer eyewall forms, usually over a period of roughly 12 hours. The second eyewall is larger than the first and thus tends to be associated with a larger footprint. However, the new eyewall will have somewhat higher central pressure and usually somewhat slower winds than the original one but may re-intensify in time.

Retrospective

While the historical storms in the era before hurricanes were named must never be forgotten, the major hurricanes of the last 25 years take on a special importance because they are fresher in the minds of many Floridians today. According to damage estimates compiled by the National Hurricane Center, seven of the top ten damage-producing storms in the United States (Katrina, Andrew, Wilma, Charley, Ivan, Frances, Jeanne) had some impact on Florida (Blake, Rappaport, and Landsea 2007). All of these hurricanes occurred in the last 25 years. The lessons these storms have taught are all for naught if we are unprepared or even underprepared for the next major hurricane.

8

Fog, Drought, and Fires

Fog

As the misty fog creeps in and thickens, it creates an eerie feeling as it envelops and shrinks our world. Fog is an often overlooked but still important component of weather and climate. In some desert parts of the world, fog provides water, the most important means of sustaining life. Desert organisms have adapted to trap fog droplets on spider webs, leaf surfaces, and even animal bodies. In more populated areas, because it can interfere with land, water, and air transportation, fog is more of a nuisance, particularly for drivers on the highways, boaters, and airline pilots. At times fog mixes with smoke from forest fires and shortens visibility to an arm's length. Most of the fog in Florida develops during the cool season.

Fog is simply a cloud at ground level. It forms for the same reason that clouds form. Whenever humidity (water vapor) near the ground is abundant, breezes are light, and the air is cooled, fog will form. This is because the full capacity of the air to allow that water to exist in vapor form has been reached and the excess water vapor must condense into liquid form. We have all experienced what happens when air cools. We have seen our humid, exhaled breath condense into a liquid cloud right before our eyes on a cold day. At very low temperatures, water vapor must convert directly into solid form, such as frost or rime ice, in a process called deposition. Sometimes the air is so moist that just moderate cooling will cause active condensation to begin. A glass of ice water may need a coaster when it is sitting on an expensive piece of furniture because the ice chills the air adjacent to the glass. This causes condensation of the water vapor in the air adjacent to the glass and a water-damage ring on the furniture if a coaster is not used.

Weather observers typically measure the temperature and the dew-point. Humidity relative to the air temperature can be calculated from those two measurements. Fog forms when the relative humidity is near 100 percent. The dewpoint is a measurement of the moisture in the air and indicates the temperature at which the air will become saturated if the air pressure is constant. Clouds in the air, fog at ground level, or dew will form when the air becomes saturated. Meteorologists have identi-fied several types of fog, distinguished by the conditions that initiate their formation.

Radiation Fog

All through the day and night, the earth's surface emits longwave ra-diation upward to the atmosphere. During the late-morning and mid-day hours, the shortwave radiant energy that the earth's surface receives from the sun often exceeds the longwave energy the earth is emitting, resulting in a temperature increase. But during the late afternoon or early evening, the amount of longwave radiant energy the earth's sur-face sends up to the atmosphere begins to exceed the shortwave radiant energy it receives from the sun. This causes the temperature to decrease. Of course, after sunset, the earth receives no solar radiant energy, and its surface cools as heat is emitted to space in the form of longwave radiation.

Radiation fog is the most common type of fog in Florida. It occurs when a net loss of radiant energy causes the air to cool to the dew point, the temperature at which active condensation begins. Radiation fog is most common on clear nights when the winds are light and the air is humid near the surface. In these conditions, the temperature doesn't need to fall too much to reach the dew point. Clear skies allow for the efficient transmission of radiant energy into space. When skies are over-cast, clouds act like a blanket by absorbing some of the radiant energy and re-radiating some of it back to the surface thereby keeping the temperature warmer than the dew point. Likewise, calm conditions allow for efficient loss of energy to space. When the winds are perfectly calm, instead of fog forming, dew may form on objects that cool faster than the air, such as automobiles or blades of grass. For dense fog to form, wind at a speed of up to 5 miles per hour must create some mixing of the lower atmosphere. Moderate winds above 5 miles per hour may lift the fog into a low layer of stratus clouds. Stronger winds interfere with the loss of longwave radiant energy from the surface by mixing warmer air aloft with the cooler air near the ground. If the moisture is in a shallow layer near the ground, a type of radiation fog known as ground fog may form; this is what we sometimes see over fields adjacent to roadways. As the air cools, it becomes denser and sinks to the lowest areas on the surface. On cold nights, the small hills in central and northern Florida have temperatures that may be initially a couple of degrees cooler than air in the adjacent valleys. As the night continues, air that cools on the hillsides becomes denser and slides downhill into the valley and accumulates there.

Advection Fog

Advection fog occurs when moist air moves laterally over a colder surface and is chilled from beneath. If the moving air is chilled to its dewpoint temperature, condensation will begin and fog will result. The lateral movement of air with a different temperature or moisture content is known as advection, so this type of fog is known as advection fog. This type of fog often has distinct edges that give the appearance of a fog bank.

What conditions produce advection fog in Florida? Florida is surrounded by the waters of the Atlantic Ocean and Gulf of Mexico. During the cool season, areas of shallow waters cool significantly. Those areas of shallow water exist along much of the Gulf Coast and along the northeast coast near Jacksonville. Advection fog forms when warm, moist air with high dewpoint temperatures is advected over the colder nearshore waters and the temperature of that air lowers to the dewpoint. Fog that forms over the saltwater areas is called sea fog. In Jacksonville and sometimes farther south, northeast winds (i.e., from the northeast) pick up moisture from the warm tropical waters that flow northward as part of the Gulf Stream and move it over the colder waters of the continental shelf, causing advection fog to form. The sea fog that is formed through this process often moves over adjacent land areas. Over the Gulf of Mexico, the shallow waters adjacent to the coast cool quickly after the first few cold fronts pass. Before the next cold front approaches, the winds become southerly (i.e., south to north) and transport warmer, moist gulf and Caribbean air northward over the cold, shallow water, where large areas of sea fog form.

Figure 8.1 shows a satellite image of sea fog over the Gulf of Mexico that is extending over land areas where radiation fog exists. Unlike radiation fog, sea fog can form day or night. Banks of sea fog sometimes move onshore during the afternoon and turn a bright sunny sky into a dark, cool, and misty blanket. The ocean is a lonely place even under the best of conditions, but imagine how terrifying it must be to navigate through mile after mile of near-zero visibility in a sea fog. It is no surprise that so many vessels have been lost at sea.

Most wind except for north winds (i.e., wind moving north to south) will usually bring humid air into Florida, either from the Atlantic Ocean,

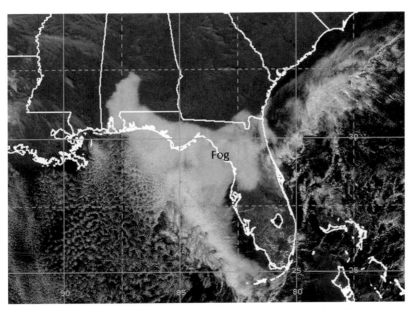

Figure 8.1. A satellite image of sea fog over the Gulf of Mexico and northern Florida. Source: NOAA Geostationary Satellite Server.

Caribbean Sea, or the Gulf of Mexico. Thus, in Florida, advection fog is most likely to occur after winds shift from the north to some other direction.

In higher latitudes and in mountainous areas, other types of advection fog are common. For example, air moving from bare land cover to a snow-covered surface is chilled from beneath and may form fog. Air moving up a mountain slope is chilled as adiabatic cooling occurs. Obviously, these types of advection fog are not common in Florida. Summer and fall are the most common seasons for fog in New England and the Appalachian region, largely because more moisture is present in the air when the air is at warmer temperatures and because summer is the season when cold air tends to meet warm air in those regions. Other areas along the Atlantic and Gulf Coasts share Florida's winter and spring fog pattern. Along the West Coast of the United States, where water temperatures are cool all year long, the fog season is skewed more toward summer, when warm, moist air masses are more likely.

Evaporation Fog

Evaporation fog forms because too much water vapor is added to the air to exist in vapor form and the moisture condenses. Two types of evaporation fog are most common in Florida. Precipitation fog forms when rain falls into cooler and slightly drier air and begins to evaporate. The evaporation saturates the air and fog forms near the surface. This can happen immediately before a cold front arrives but is more likely near a warm front. As moist air is lifted over cool air along the shallow slope of a warm front, clouds and rain develop. The rain falls into the slightly cooler drier air, evaporates, and saturates the air, forming fog. A warm front is the most likely type of mechanism to produce precipitation fog because it is associated with light precipitation. Intense precipitation would have less time to evaporate before hitting the surface and would be less likely to form fog. Dreary days with precipitation fog are more likely in the colder winter air over northern Florida and the Panhandle.

A second type of evaporation-induced fog occurs when cold air moves over a warm body of water such as a swamp or a lake. The air adjacent to the water rises as the water warms it because it is less dense than the colder air overlying it. As this moist air rises and begins to cool, some of the liquid water in it evaporates. If the evaporated water causes the adjacent cooler air to become saturated, fog results. This often happens in hilly areas where radiational cooling creates denser down-sloping air that moves over streams, lakes, swamps, or runoff from irrigated farmland and creates evaporation fog. This type of advective evaporation fog is called valley fog. That is why during the cool season, on clear nights when dewpoint temperatures are in the 50s and 60s, we often see fog forming first over lower areas in the terrain and over lakes and rivers. At times, the evaporation fog is less dense with thin, wispy columns of fog rising over an inland body of water reminiscent of steam rising from a pot. This is called "steam fog." The term is misleading because no true steam is involved. Steam fog is most common in the fall, when temperatures in water bodies are still quite warm from the long Florida summer and when much colder polar air masses begin to make their way southward into Florida. The eerie fogs Floridians sometimes see at Halloween are often seasonal steam fogs.

Fog Climatology of Florida

According to a report for the Florida Department of Transportation (Ray, Du, and Rivard 2014), Miami and the southeastern part of Florida have only a few fog days per year (map 8.1). From Miami, the number of days with fog increases northward through the state. The foggiest part of the state is over northern Florida. The area near Tallahassee has up to 55 days per year of fog. Little fog forms during the warm season, when radiational cooling is limited and what does form is typically shallow ground fog. The Gulf of Mexico coastal areas are notorious for huge expanses of sea fog that develop and push onshore when the winds are from the south. Northern Florida and coastal areas along the west coast north of Tampa Bay are also influenced by overnight radiational cooling that produces widespread fog over land areas. Along the east coast, sea fog is most common from near Jacksonville to Cape Canaveral as northeast winds pick up moisture and warmth over the Gulf Stream that is then cooled by the nearshore waters and condenses into sea fog. The many small lakes that dot Florida's landscape, such as those formed by sinkholes, can have abundant steam fog nearby that is too localized to appear on any state map of fog frequency.

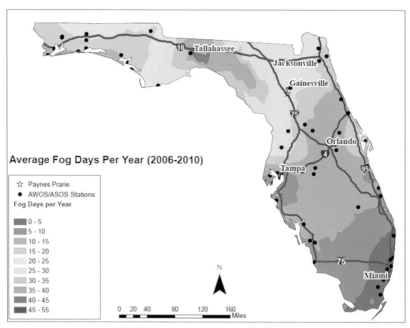

Map 8.1. Average annual number of days of fog over Florida. Source: Ray et al. 2014.

Droughts in Florida

When the rains are sparse or absent for weeks or months and the sun continues to beat down on Florida, stream and lake beds get dry and the ground begins to crack. During these drought conditions, sunny days become an unwelcome sight for residents. Drought creeps in silently, often becoming a problem before the general public realizes that it has begun. Droughts are a part of Florida's climatic history, and with a growing population that now exceeds 20 million people, water shortages have become a serious problem. Droughts usually begin as the summer rainy season comes to an end, often as the first strong cold front of the upcoming cool season ushers drier air over the state. When the dry area of high pressure behind the front lingers and prevents moisture from coming back over the state, drought begins to take hold, especially if conditions behind the front remain relatively warm. If the dry pattern stretches into months, serious water shortages are likely. Drier winters are often associated with La Niña.

Figure 8.2 shows how much annual rainfall accumulations can vary. Notice that Tampa's rainfall was at its lowest—only 29 inches—in 1956 and the next year it was up at record levels to near 70 inches. That record high amount of rainfall was shattered when Tampa received 77 inches of rainfall in 1959. During the strong El Niño of 1997, Tampa had 68 inches of rain, but as the dry conditions of La Niña set in, rainfall dropped to only 30 inches in 2000. It is difficult to develop a sensible, sustainable water use plan for so many people when supplies are so erratic and difficult to anticipate.

An extreme example of how susceptible Florida is to drought occurred in 1998. The winter of 1997–1998 was dominated by an El Niño pattern in which the frequency and amount of rainfall was far above normal. Some areas of the state had two to three times the normal winter rainfall. The rains ended abruptly around the beginning of March 1998, when the El Niño flip-flopped into a La Niña and the state began to dry out. No rain fell in March, and the same thing happened in April as temperatures climbed under cloudless skies. Drought had begun.

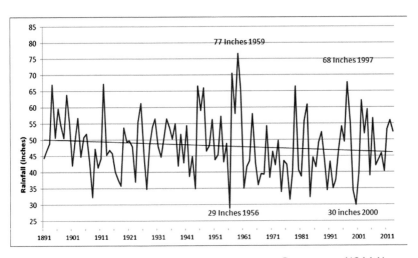

Figure 8.2. Annual rainfall variation at Tampa, 1891–2014. Data source: NOAA National Centers for Environmental Information.

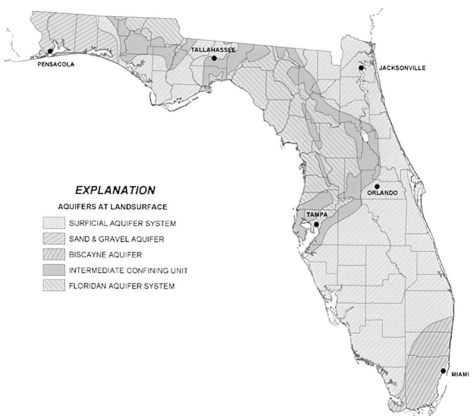

Map 8.2. Major aquifers of Florida. Source: Florida Department of Environmental Protection 2015.

Sinkholes

Drought brings about another nagging problem across a large portion of the state. The land of Florida consists of mostly sandy soils on top of sedimentary limestone composed largely of calcium carbonates. Over time, the limestone dissolves in places as the result of slightly acidic rain water. A limestone foundation with large underground spaces that become underground drainage systems is known as a karst landscape.

These drainage systems below the ground are aquifers that supply most of the drinking water Floridians consume. Several aquifers flow slowly underground through the state. The most prominent is the Floridan Aquifer System, which extends from northern Florida along the Gulf Coast to Tampa Bay and deepens to 3,000 feet as it flows southward

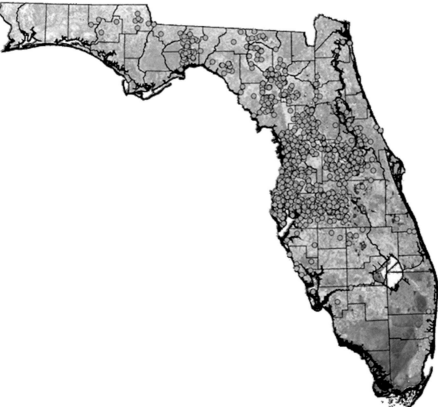

Map 8.3. Locations where sinkholes are prevalent in Florida. Source: Florida Geological Survey.

(map 8.2). Other aquifers are the Sand-and-Gravel Aquifer under the western Panhandle; the Intermediate Aquifer System, which supplies water to southwest Florida counties; the Surficial Aquifer System, which surrounds the Intermediate Aquifer System and covers much of Florida; and the Biscayne Aquifer under southeast Florida, which provides drinking water to Miami-Dade and Broward Counties, centers of high population.

When the aquifers are full, they support the land on top, but when rainfall is scarce and aquifer levels are lower, the land is not supported by the water and may fall into an underground cavern. This is how sinkholes are formed. Many sinkholes eventually become lakes (map 8.3). Most of the sinkholes form within the deeper Floridan Aquifer System.

Sinkholes are a prominent news feature in Florida when homes and automobiles are gobbled up by the collapsing karst landscape.

Florida Wildfires

Although Florida has abundant precipitation by world standards, it is still vulnerable to periods of drought and wildfires because of the annual pattern of rainfall distribution. According to the Florida Forest Service, over 5,500 wildfires occur annually in Florida, burning over 218,00 acres (Florida Department of Agriculture and Consumer Services, Division of Forestry 2011). The three ingredients for fire are heat, fuel, and oxygen. Florida's wildfire season occurs during the cool season after the almost daily summer thunderstorms end. As the dead branches and leaves from trees, shrubs, brush, and grasses dry out, fires become more likely. The height of the fire season is during the spring, before the summer rainy season starts. The few lightning storms of spring bring little rain but provide the spark that sets the fiery process in motion. Figure 8.3 shows smoke from fires over north Florida and south Florida that was transported over the Gulf of Mexico and Atlantic Ocean.

In addition to dead branches, leaves, and grasses that easily catch fire, Florida has some unique vegetation that contains combustible oils that will burn even while the plant is alive. These flammable plants include galberry, wiregrass, saw palmetto, cabbage palm, and pine trees. Most of the wildfires in Florida are surface fires that burn fuels near the ground. Almost half of the state's surface fires are fed by grass fuels. About a third of the fires occur in palmetto-galberry areas. Occasionally fires climb into the crown or tree canopy, burning moist vegetation as they are fanned by wind, but this is rare. Some fires start on the surface and ignite parts of a dry swamp that contains damp peat and logs. Once these ground fires are started, they burn slowly and emit considerable smoke. Some may burn for weeks.

A third of the state's wildfires are ignited by lightning. Unfortunately, human activities are responsible for the other two-thirds of wildfires. Arson, carelessly tossed cigarettes, campfires, burn piles, and vehicles are the cause of many wildfires in Florida. It doesn't help that so many

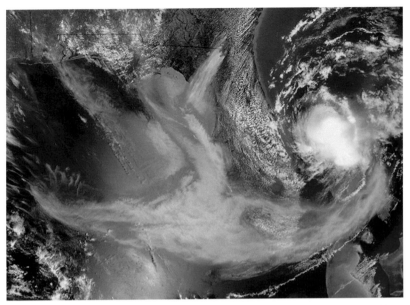

Figure 8.3. Smoke from fires over North Florida and South Florida. Source: NASA.

people drive near and live close to vegetated regions that can become tinderboxes.

Many Florida plant species require occasional fire to sustain life by leaving open soil areas for seed growth and providing new light for new growth and species that could increase the ecosystem's biodiversity. Foresters regulate the amount of fuel in forested areas by igniting controlled fires over limited forest areas to reduce the threat of out-of-control wildfires. Prescribed burns are common in Florida and occur on a regular basis, particularly during the dry season. Considerable care is taken to burn on favorable wind days to keep smoke away from roads and from the wildland-urban interface, where homes are vulnerable.

The 1998 Wildfires

Wildfires ignited all over Florida during 1998. Oddly, that year, the winter brought copious rain from El Niño that was measured in feet. In March, with the onset of La Niña, the rain stopped abruptly. Temperatures rose, the state dried out, and the stage was set. In May, highly electrified thunderstorms with very little rain developed along the sea breezes and set the state on fire. A FEMA report documented the billion-dollar impact of 2,200 individual fires that burned nearly 500,000 acres. Homes and cars also went up in flames. Over 10,000 firefighters courageously fought the blazes from May through June. The fires were not extinguished until July, when the rainy season finally began.

Smoke and Fog: A Deadly Combination

When the relative humidity is high and fog begins to form near a smoldering burn area, visibility can go to zero. The combination of smoke and fog drifting over roadways has created many highway disasters. Obviously, Florida would be a location that is susceptible to this hazard. The fog mixes with the damp, smoldering organic material that adds fine particulates and heated water vapor to the air. This provides a huge reservoir of particles for water vapor to cling to as it condenses into fog. The mixture of smoke and thick fog drifts through low terrain when winds are light. This can create a veritable blanket of fog and smoke, reducing visibility to zero instantaneously as the deadly mix drifts over roadways. In addition to the sharply increased hazard of road accidents, breathing such a concoction is hazardous to respiratory systems of nearby people and animals, especially when vehicle exhaust from the likely traffic jams near the fire enters the mix.

A horrific accident occurred in the early morning hours of January 9, 2008, when thick smoke from a prescribed burn that had become a wildfire the previous day drifted across a narrow stretch of Interstate Highway 4 (I-4) in a remote area of Polk County in central Florida. Collins, Williams, and colleagues (2009) reported that as drivers entered the blanket of thick smoke and fog, visibility instantly dropped to zero and 70 cars and trucks collided within a short time. Five people died and 38 were injured during the pileup. A few years later, on January 29, 2012, a similar traffic accident occurred on I-75 in Alachua County near Gainesville. Eleven people died in that accident and 46 others were injured. After these accidents, the Florida Highway Patrol became much more proactive about closing portions of highways when the possibility of smoke and fog could limit visibility for drivers.

Good Weather Can Be Bad Weather

During the tourist season when the weather is calm and cool and hurricanes and lightning are not a threat, meteorologists are still busy, forecasting hazardous phenomena that occur on fair-weather days. Fog and smoke hazards reduce visibility that decreases reaction time for drivers and operators of vessels. Droughts and wildfires pose an inherent danger to people's homes, lives, and livelihoods. Most of Florida's water comes from the precious aquifer system that needs consistent rainfall. The cyclical nature of rainfall from one year to another requires robust engineering to drain water when too much falls so it can be stored for times of drought. Even as Florida's population increased, few manmade reservoirs were built because of Florida's flat terrain. The challenges of co-existing with the whims of Mother Nature in a densely populated and still-growing state can only increase in the future.

9

Florida's Coastline
and Beaches

Florida's unique coastline is known for its beautiful white sand beaches and sparkling water areas. Tourists and residents often seek a sandy shore where they can plunge into the surf or cast a fishing line. With 1,197 miles of water surrounding the state, Florida has more coastline than any other state except Alaska. Most of the coastline is comprised of either sandy beaches, salt marsh, or coral reef. Florida's beautiful sandy beaches account for 663 miles of coastline. Weather and climate continuously affect the coastline and are affected by the coastal zone.

The Coastline and Barrier Islands

The barrier islands along much of the Florida coastline are naturally formed by waves and currents that push available sand toward the shore. The sand builds up, first as an offshore sandbar under water, then protruding out of the water. Eventually the newly formed barrier island supports vegetation, which in turn stabilizes the landform and makes it less vulnerable to erosion from precipitation, storms, and wave action. The shape of Florida's beaches changes by season. During the summer, when waves and currents are weaker, the beaches have a gentle slope and prominent sand bars are aligned parallel to them. During the winter, the more frequent stronger waves and currents spread the sand offshore and create a steeper beach slope. Also during the winter, when

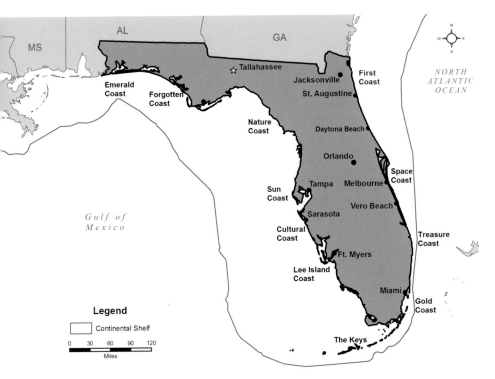

Map 9.1. Nicknames of locations on the Florida coastline and location of the continental shelf.

cold fronts pass and strong north winds blow, longshore currents develop and transport sand along the beaches from north to south in the peninsular region of Florida. Barrier islands are not continuous around the state and form only where the water is deeper offshore and where wave action is significant enough to move the sand.

Florida's coastal areas have been assigned nicknames related to their physical or historical characteristics (map 9.1). The barrier islands in the Florida Panhandle stretch from the Alabama-Florida border to Dog Island (near Apalachicola) and have some of the whitest and finest sand in the world. The stretch of coastline with emerald-colored waters along the western Panhandle is known as the Emerald Coast. The less-populated area of the Panhandle near Apalachicola is known as the Forgotten Coast. Apalachee Bay, also known as the Big Bend area, is located east of Dog Island, where the beaches fade to salt marsh and large waves become nonexistent in the extensive shallow waters of the broad continental shelf. The southern part of the Big Bend area is also known as the Nature Coast. Seahorse Key near Cedar Key is a sand-dune island that developed more than 12,000 years ago, when glaciers covered much of North America. Its height of 52 feet makes it the highest point on the west coast of Florida. The southwest coast of Florida stretches from Anclote Key (near Tampa) south through the Tampa Bay area. The Tampa Bay area is known as the Sun Coast. The barrier islands along the southwest coast continue south through the Sarasota area, which is known as the Cultural Coast, and the Lee Island Coast, which encompasses the greater Fort Myers area, including Sanibel Island south to Marco Island. The Florida Keys, which are comprised of coral reef, jut out from the mainland at the southern tip of the state. The Keys stretch for 127 miles and are sandwiched between the shallow waters of Florida Bay and the Florida Straits. The sandy barrier islands resume along the southeast Florida coastline, known as the Gold Coast, which stretches from Key Largo through Miami to Fort Lauderdale. The Treasure Coast stretches north from Boca Raton to Vero Beach. Most ocean swells, except north swells, which come into the southeast coast from the Atlantic Ocean, are blocked by the islands of the Bahamas. The barrier islands continue northward along Florida's east coast, where the beaches are exposed to the open Atlantic Ocean. The area known as the Space Coast stretches north from Melbourne through Daytona Beach. The First Coast stretches north through St. Augustine and Jacksonville to the Florida/Georgia border. Each of these segments is characterized by slightly different weather and climate settings that contribute to the uniqueness of their landscapes.

The Continental Shelf

The location and extent of the continental shelf are important from a weather and climate perspective because the shallow waters over the shelf warm faster toward summer and cool faster toward winter than the deeper waters off the continental shelf. The edge of the shelf is often a zone where air of significantly different temperatures can interact, especially where tropical deep water is adjacent to the winter-cooled water of the shelf. Along the Florida coast, these temperature differences are not nearly as significant as they are farther north along the Atlantic coast, but they can still be significant enough to have a significant impact on the weather. And finally, the frictional effects of a longer extent of shallow shelf waters decreases the intensity of powerful waves that develop during storms in the open oceanic waters.

Florida's continental shelf and the underlying limestone are shaped by draining rains, flowing rivers, and crashing waves. Many rivers feed into the Gulf of Mexico, but except for small creeks, only the northward-flowing St. Johns River feeds into the Atlantic Ocean. The Florida Gulf Coast (map 9.1) has a broad continental shelf that extends out 150 miles, except along the Panhandle area, where it extends only about 50 miles offshore in areas. The depth of the Gulf of Mexico plummets from 300 feet at the edge of the continental shelf to about 10,500 feet in the deep waters of the central gulf. Along the east coast, the continental shelf is close to the coast near Miami but gradually extends farther offshore from around Vero Beach northward.

Weather and Ocean Influences

Florida's weather is influenced by the surrounding Atlantic Ocean and the Gulf of Mexico. Large and small ocean currents are at work along the coasts. In chapter 3, the idealized view of climate across the globe showed that warm air in the tropics adjacent to the equator rises and is transported north and then sinks around 30°N. The sinking motion of the air, or subsidence, and accompanying adiabatic drying of the air creates the semi-permanent areas of subtropical high pressure around the globe that support many of the great deserts of the world. The subtropical high-pressure areas also influence large parts of the world's ocean basins. In the Northern Hemisphere, the clockwise and outward surface wind flows around the subtropical high-pressure areas support the northeasterly trade winds. On the north side of the subtropical high-pressure areas, the clockwise and outward wind flows have earned the name the westerlies because they move from west to east. The scenario is similar over the Southern Hemisphere. These large consistent areas of high pressure help drive the great curving ocean currents, or gyres, which are also

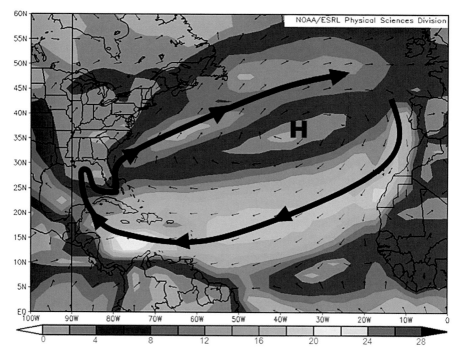

Map 9.2. Average annual wind direction and speed (in mph) associated with the Bermuda-Azores high and associated surface ocean currents. Source: NOAA Earth System Research Laboratory (n.d.)

influenced by the earth's rotation and the continents. Gyres are large circular surface currents associated with large semi-stationary areas of high pressure. The major gyres rotate clockwise (Northern Hemisphere) and counterclockwise (Southern Hemisphere).

Ocean Currents

The distinct semi-permanent area of high pressure over the Atlantic Ocean called the Bermuda-Azores high drives ocean currents that influence areas near Florida. Their influence also extends as far away as Europe. Map 9.2 shows the average annual wind pattern associated with the Bermuda-Azores high and the associated water movement. The dominant northerly winds along the coast from southern Europe to North Africa turn and become a long stretch of northeasterly trade winds.

These persistent trade winds push Atlantic Ocean water from Africa westward into the warm Caribbean Sea and toward Central America. This broad flow of warm water has only one escape route out of the Caribbean Sea (map 9.3a). The waters squeeze northward and through the narrow Yucatán Channel passage between Cuba and Mexico. From that point, the narrowness of the channel causes the current to accelerate (much like water shooting out of a hose nozzle) and shoots northward into the deeper waters of the Gulf of Mexico. The Gulf of Mexico also offers only one escape route for the built-up momentum of the current, and that forces the current to loop back southward. Because of its characteristic path, this current in the Gulf of Mexico is named the Loop Current. The current then becomes the Florida Current as it exits the Gulf of Mexico through the deep water of the Florida Straits between Florida and Cuba. The Florida Current races northward between Florida and the Bahamas as it is joined by the Antilles Current, which flows westward between Cuba and the Bahamas. This large, swift, and powerful circulation pattern becomes known as the Gulf Stream as it takes warm, tropical water northward through deep waters, past Florida, across the Atlantic Ocean to Europe.

The Loop Current has a cycle in which the northernmost part of the loop closes off and drifts westward toward Texas. This leaves a much shorter loop that eventually grows northward and closes off once again. The length of the Loop Current cycle varies from several months to over a year and is dependent on the winds and the current flowing into it.

The currents that move through the waters past Florida are a source of energy that strengthens storm systems as they move through the Gulf of Mexico. Hurricane Katrina's intensification over the Loop Current in 2005 provides a great example of what can happen in such interactions between the ocean and the atmosphere. During the cool season, the warm waters of the Loop Current often provide heat energy that strengthens thunderstorms that precede cold fronts charging eastward toward Florida. But often the impact is only temporary because the cool nearshore waters drain off the thunderstorm's strength.

Thunderstorms moving offshore from Florida's east coast can intensify explosively over the Gulf Stream, which can lead to large waterspouts and hailstorms. This may be a reason for the mysterious disappearance

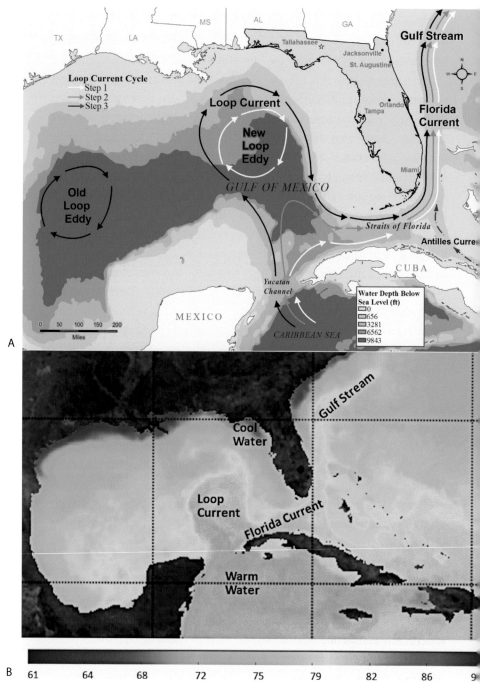

Map 9.3. *A*, Ocean currents around Florida; *B*, Surface water temperatures along the coast of Florida (°F) in January 2016. Source: NASA Short-Term Prediction Research and Transition Center.

of boats and airplanes within the mystical Bermuda Triangle. The major currents and the more local wind-driven currents that move around Florida have been known to carry those who unwisely venture out in poorly equipped boats. Once a boat capsizes and the occupants are in the water, it's a fight for survival against the elements, and the human body is notoriously poorly equipped to survive in such situations. Sharks are much less of a problem than dehydration, sunstroke, and debilitating muscle cramps from bobbing in the water while clinging to a boat. One case that made both national and international headlines was detailed in the book *Not without Hope* (Schuyler and Longman 2010). In February 2009, four football players went on a fishing trip in the Gulf of Mexico. Marquis Cooper and Corey Smith played in the National Football League for the Oakland Raiders and the Detroit Lions, respectively. Will Bleakley and Nick Schuyler were football players for the University of South Florida. Their boat anchor became stuck and their boat capsized after they made the mistake of tying the anchor rope to the stern of the boat and hitting the throttle. The athletes suffered hunger, dehydration, hypothermia, and hallucinations and their bodies were battered by the huge waves. After more than forty hours, Schuyler was the only one found alive. The Coast Guard had a difficult time locating them in the bad weather; they had difficulty discerning the boat amid the white crests of the waves and the currents that moved them around.

Water Temperatures

Thinking of taking a dip in the ocean or gulf in January? Depending on where you are, you might want to think twice! Although the deep waters around Florida channel the warm Gulf Stream, shallow waters heat and cool faster than deep waters. Cold continental runoff from streams also promotes the rapid cooling of coastal waters in winter and the strong thermal gradients in estuaries. Map 9.3b shows that average January ocean temperatures drop to the mid-50s in the Panhandle Gulf waters near Pensacola. The same is true on the Atlantic side in the areas of northeast Florida, near Jacksonville and St. Augustine, where Florida's coastline curves slightly westward, away from the warm waters of the Gulf Stream. Water temperatures drop to the mid-50s along this shallow

Table 9.1. Average ocean and gulf water temperatures around Florida

Location	Jan.	Feb.	March	Apr.	May	June	July	Aug.	Sept.	Oct.	Nov.	D
Fernandina Beach	55	55	62	72	78	81	84	84	81	72	66	
Mayport	58	58	62	72	78	81	83	83	82	75	69	
St. Augustine Beach	57	56	61	71	77	81	84	83	82	72	67	
Daytona Beach	61	59	65	73	78	80	80	81	82	76	71	
Stuart Beach	67	66	70	74	78	79	79	80	80	77	75	
Miami Beach	71	73	75	78	81	85	86	84	83	79	76	
Virginia Key	71	72	74	78	83	85	87	86	85	81	76	
Key West	69	70	75	78	82	85	87	87	86	82	76	
Naples	66	66	71	77	82	86	87	87	86	81	73	
St. Petersburg	62	64	68	74	80	84	86	86	84	78	70	
Cedar Key	58	60	66	73	80	84	86	86	83	76	66	
Pensacola	56	58	63	71	78	84	85	86	82	74	65	

stretch of coastal waters during the winter. Farther south, near Miami and in the Florida Keys, average water temperatures in January exceed 70°F. In the summer, average water temperatures are more consistent statewide, rising steadily to a peak from July to September. This peak ranges from the low to mid-80s near St. Augustine to the upper 80s and sometimes more than 90 along the coastline of southern Florida.

During the summer, the Gulf of Mexico and Atlantic waters warm more slowly than the land, especially where the water is deeper. The warmest water temperatures for the year occur during late summer, when the land areas have already begun to cool (Table 9.1). The cool season is also lagged over water; the coldest land temperatures occur in December and January and the coldest water temperatures occur slightly later. This is because of water's conservative nature; it gains heat slowly in the summer and releases heat slowly in the winter. As an example, the coldest water temperature off the coast of St. Augustine occurs in February and the warmest water temperature occurs in September, toward the end of summer.

Localized currents can affect water temperatures substantially along the beaches. When persistent winds blow across coastal water areas, this creates a common phenomenon called Ekman flow, in which the net water transport is at an angle to the right of the wind flow (in the Northern Hemisphere) and to the left of the wind flow (in the Southern

Hemisphere). Ekman flow is essentially the Coriolis effect within the ocean, originating from the rotation of the earth. In addition, the deflection decreases with decreasing latitude, so it is slightly less substantial in Key West than it is in Pensacola.

When northerly winds blow along the west coast of Florida, the top layer of water moves offshore to the right of the wind flow, or from east to west. As the top layer of water moves away from the coast, cold water upwells from underneath to replace the top layer of water. This explains why water temperatures plummet along beaches in the Tampa Bay area when the first strong cold fronts of the season pass through.

The opposite is true along Florida's east coast. When warm southerly winds blow along the east coast, this moves the top layer of water offshore and creates the cold upwelling scenario that can drop water temperatures in the summer from the mid-80s to 70°F. But Ekman flow associated with northerly winds along Florida's east coast causes an influx of warm water from the Gulf Stream on Florida's shores, keeping water temperatures warmer.

Wave Formation

Ocean waves are simply energy moving through water. They are usually generated by the wind, but earthquakes, landslides, and other seismic activity can also create waves known as tsunami (the plural of "tsunami" is also "tsunami"), which means "harbor wave" in Japanese.

Waves that are common along Florida beaches get their start as winds blow over open ocean areas. The wave generation region is known as the fetch. Stronger winds over longer ocean fetches grow larger waves. Large winter storms and strong hurricanes that produce strong winds over long fetches of ocean create the highest waves. Waves are measured in several different ways (figure 9.1). The height of a wave is the vertical distance between the trough and the crest of the wave. The length of a wave is the distance between each successive crest or trough. The period of a wave is the time it takes for a complete wave to pass a point; it is measured in seconds. The longer the period between waves, the longer the waves are and the faster the waves move when in deep water. Waves generated by the wind vary in height and period. A wave with a short

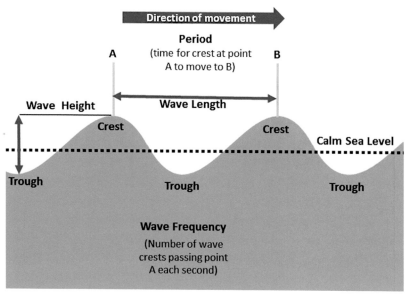

Figure 9.1. Wave measurements.

period of four seconds moves through deep water at 12 nautical miles per hour (knots), while a longer-period wave of 20 seconds moves at 60 knots. These faster long-period waves move out of the fetch areas in groups. When a distant storm has created longer-period waves, the waves will arrive at beaches in groups or sets and there will be several minutes between each set of larger waves. As waves enter shallower water, shoaling takes place. This is when the deepest part of the wave is affected by the bottom and slows while the top continues to move faster and steepens. Waves will break when the water depth is about 1.7 times the wave height.

Buoys

Wave characteristics are determined from buoy data. Oceanographic buoys have instruments that measure the varying waves, which may be distant swells coming from different directions superimposed on locally generated wind waves. The information is analyzed by a computer algorithm that sorts out and categorizes the waves. The significant wave height, reported by buoys, is calculated by averaging the highest third of the waves. However, the highest waves may be twice as high as the significant wave height. When a 10-foot significant wave height is forecast, some of the waves may occasionally be up to 20 feet high. When waves merge, the resulting wave can be much larger and may be classified as a rogue wave if the seas are already large.

Tsunami

The last major tsunami to strike Florida, which was estimated at around 10 feet, was spawned across the Atlantic Ocean by the Great Lisbon earthquake of 1755. Tsunami can be devastating because they are long waves that can race up to 500 mph across the deep ocean, leaving those in its path vulnerable to the wave's shocking arrival. Once the wave approaches shallower water, the water typically retreats far from the coast and then comes in surging, rising higher than anyone might expect.

Sometimes tsunami-like waves develop from rapid changes in atmospheric pressure moving quickly over the shallow continental shelf areas offshore. These waves are often small, but the first well-documented case in Florida occurred in 1992, when a large wave came out of the dark, and swamped cars parked on Daytona Beach at night. A few years later, another wave moving over 50 miles per hour impacted the Florida west coast from the Tampa Bay area southward to Naples; that wave was estimated to be ten feet high. These waves, known as meteotsunami, were produced by relatively sudden and substantial changes in atmospheric pressure.

Tides

Tides are also considered to be waves. They are created by the gravitational effect that pulls the waters toward the moon and the sun. The larger but more distant sun's gravitational effect on the tides is about half that of the moon. The highest and lowest tides are created during the full and new phases of the moon when the gravitational effect of the sun and the moon work in unison. Incoming tides are called flood currents and outgoing tidal currents are called ebb currents. The shape and orientation of the coastline, the bathymetry (the depth profile of the waters), and other factors besides the lunar phase and position relative to the sun determine the daily range of tides at a location. Locations in the Bay of Fundy in Nova Scotia experience daily tidal ranges over 35 feet, and many locations around the world experience tidal ranges over 20 feet. Most of Florida's coastline, however experiences relatively small changes between high and low tides of 2 to 4 feet. The tidal cycle between high and low tides is about 6 hours and 12.5 minutes. Thus, because there are four tidal cycles in a 24-hour period, each day the tides occur 50 minutes later than the day before. Along most of Florida's Atlantic coast, two high tides of nearly the same heights and two low tides of nearly the same heights occur during the almost daily cycle. Along the gulf coast, the heights of tides vary with major and minor high and low tides, and some days have only one high and low tide.

Beach Safety

Given the extensive coastline, it's not surprising that more people drown in the ocean in Florida than any other state. The ocean can look like a pool or a lake, but looks can be deceiving. When enjoying the magnificent beaches of Florida, it is important to follow several safety guidelines.

First, prepare before the beach visit by checking the weather and wave forecasts. Several websites monitor the waves and forecast surf conditions days in advance. Second, especially for those who rarely visit the beach, it is important to find a beach with a lifeguard. This is particularly important for those who want to splash in the water but cannot swim. Third, it is important to be aware of where the most dangerous ocean conditions are likely to occur. Sandbars aligned perpendicular to shore often have deep areas between them. As incoming waves rush over the sandbars and up onto the sand, the water retreats back into the ocean as a rip current. Rip currents accelerate through breaks in the sandbars that allow the water to funnel rapidly into the offshore waters beyond the breaking waves (figure 9.2a). Figure 9.2b shows what the rip current channels look like when they are not covered by water.

Beachgoers should be aware of the behavior of water in these more dangerous, deeper areas that often look safer because waves may not be breaking there. Don't be mistaken, though: rip currents can pull beachgoers out into deeper water and can put even the strongest swimmers in peril if they try to fight the current. When being pulled out to sea by a rip current, swim away perpendicular to the current toward a

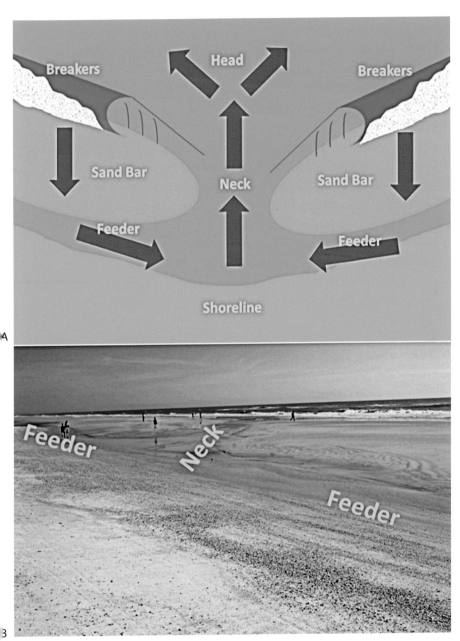

A

B

Figure 9.2. *A*, Diagram of a rip current; *B*, A rip current feeder and neck channels at low tide. Credit: Charles H. Paxton.

shallow sandbar area where waves are breaking. If you can't swim out of the rip current, tread water or float until the rip current dissipates past the breakers. Rip currents often occur adjacent to piers and jetties. If a longshore current is pushing toward a jetty, the longshore current will become an outward-moving rip current near the jetty. It is best to stay far from these manmade structures because they increase the chances of rip currents.

Finally, one of the most important guidelines to follow when in the ocean is to know how to swim and how to float and tread water. Even three-foot waves can be powerful when breaking in shallow water, and rip currents can occur with waves even smaller. If you notice that someone is struggling in the water, call 911, then grab a flotation device such as a throw ring, surfboard, boogie board, or even a plastic cooler. Rescuers often become the victim while trying to rescue a panicking person in the water. Lifeguards carry floats when on patrol, not to help them swim but to toss to a panicking person in the water. Enjoy the beaches and be safe!

10

The Changing Climate

Over 66 million years ago, dinosaurs roamed an earth that was 20°F warmer than today. All it took was a comet or asteroid of perhaps 6 miles in diameter plummeting through the atmosphere to annihilate much of the life on earth. The impact sent a shock wave around the globe, sparked massive fires, and shot dust and debris into the atmosphere that reflected and absorbed incoming sunlight, preventing much of it from warming the earth's surface. It didn't take long for the abrupt decrease in solar radiation to cool the earth. The life forms that could readily adapt to a rapidly cooling earth survived, but those that couldn't perished, including the reptilian dinosaurs and most other large animal species. As the dust settled, the planet again warmed and life evolved.

Temperatures through Time

The earth formed 4.6 billion years ago. Its climate has been quite variable, from at least 2.5 billion years ago, when temperatures were considerably warmer and colder than our current temperatures. The amount of sunlight reaching Earth varies over long time scales, resulting in ice ages and warmer periods. Did you know that we are still in an ice age? Ice ages last for millions of years, and the warm periods in between last even longer. Our current ice age began about 2 million years ago. The temperatures have varied considerably during the ice age we are currently in. There have been cycles with about 80,000 years of cold glaciation and about 20,000 years of interglacial warmth within this ice age. We are experiencing the interglacial warmth now, with temperatures about 20°F warmer than during a glaciation. An ice age is any time in Earth's history which ice persists throughout multiple years (summers and winters) somewhere on Earth. Today's Earth has "permanent" ice over Antarctica, Greenland, and the tops of the highest mountains, so we are in an ice age.

Florida's Changing Coastline

Imagine that instead of oppressive 95°F heat and humidity during Florida's summer, the high temperatures typically only reached 75°F. This sounds great, except that the flip side is that winter temperatures would regularly plunge below freezing in Miami, and north Florida would occasionally sink below zero. What could be worse? We will get to that.

The earth has a fixed water supply. It receives little from outer space and does not lose any appreciable volume. The water frozen in glaciers during cold periods of glacial advances leaves significantly less water available for the oceans, so the sea level can be as much as 300 feet lower during glacial advances than during interglacial warm periods. This leaves vast undersea areas of the continental shelf exposed during glacial advances. The land area is therefore much larger. The dashed line in map 10.1 indicates one estimate of the different shape of the Florida Peninsula during a glacial advance. The shape of Florida looks much different during glacial advances in ice ages when the global temperatures are cold.

Today, the continental shelf that extends into the Atlantic Ocean and the Gulf of Mexico varies in length outward from the shore around the state. Along Florida's southeast coast, the shelf is narrow. Even with much lower water levels, the Miami-area coastline might have been similar to the current configuration during a glacial advance. Other city areas would have been much different from today during the peak of the last glacial advance, about 18,000 years ago. Tampa Bay would have been nonexistent or possibly a large lake with the gulf coastline extended more than 100 miles away to the west. The Florida Keys would have been part of the mainland.

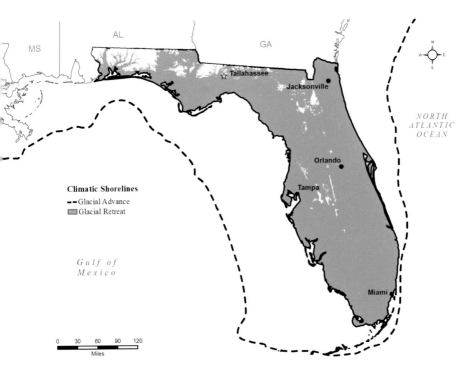

Map 10.1. Florida's changing coastline. The dashed line indicates the approximate coastline during the previous glaciation. The white areas within the state represent what the current coastline might look like under a glacial retreat scenario if all of the ice on the earth melts. Source: Adapted from Florida State Geological Survey 1994.

Florida is flooded during warm (non–ice age) periods and during intense interglacial phases within ice ages that have less glacial ice and higher sea levels than we are seeing today. During past warm interglacial periods, the Florida Peninsula was almost covered by water from sea level rise associated with the thermal expansion of the oceans and with the melting of glaciers and continental ice sheets. Currently, we are in an interglacial warm period within this ice age, but two and a half million years ago (i.e., before the present ice age), the climate was even warmer and Florida was mostly a series of islands in a sea that was over 200 feet higher than today (see map 10.1).

Natural Climate Variations

Several natural factors account for the cycles of glaciation and interglacial periods the earth has experienced. In the early 1900s, the Serbian
astronomer Milutin Milanković discovered three solar cycles that affect
climate; the Milankovitch cycles are named for him. When these cycles
are synchronized, the earth is likely to be at its warm or cold extremes.

The earth's orbit around the sun is not perfectly circular; it is slightly
oval–shaped. The amount of deviation from a true circle slowly changes.
On an approximately 95,000-year cycle, the earth's orbit around the sun
varies from within 1 percent of being perfectly circular to a maximum
eccentricity of 11 percent and then returns to within 1 percent of a perfectly circular orbit. With this oval orbital shape, Earth is slightly closer
to the sun twice during the complete cycle, or every 24,000 years.

On a 41,000-year cycle, the tilt of the earth's polar axis changes too,
from 21.4° to 24.5° and back to 21.4° again. This means that the Tropic of
Cancer and Tropic of Capricorn are sometimes more or less than 23.5° of
latitude from the equator. When the Tropic of Cancer and the Tropic of
Capricorn are farther from the equator, the direct rays of the sun extend
farther poleward in the summer hemisphere, making summers outside
the tropics warmer than they are today. In addition, the sun's direct rays
are farther away in winter than they are today, making winters colder.

In addition to these two cycles, the earth's axis wobbles like a top on a
23,000-year cycle. This means that the North Star, Polaris, is not always
aligned directly with north. During these cycles, the earth sometimes
receives slightly more or slightly less than average solar radiation, resulting in a warmer or a cooler Earth.

The Evidence of Past Climate Change

Scientists use tree rings, corals, caves, sediment cores from the deep sea,
ice cores, boreholes, and other sources to provide an estimate of actual
temperatures. These proxy techniques show changes in the earth's temperature as far back as 5 million years. In caves, stalactites and stalag-
mites provide evidence of past climates, just as tree rings do. The air

Solar Insolation, Temperature Cycle, and Atmospheric CO2

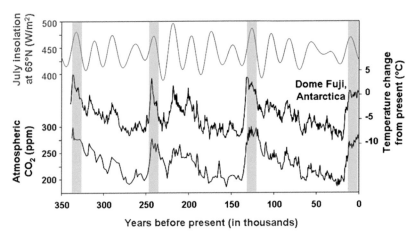

Figure 10.1. July insolation (watts per square meter), temperature change (°C), and carbon dioxide (CO_2, parts per million) going back 350,000 years from the present time, from Antarctic ice cores going back 350,000 years. Gray bars indicate interglacial warm periods. Source: NOAA National Climatic Data Center (2008).

bubbles trapped in samples of ice cores from Antarctica and Greenland and from alpine glaciers give scientists an amazing look at the past. Temperature data locked in Antarctica's ice near Vostok (figure 10.1) allow scientists to estimate the amount of solar insolation, temperature levels, and carbon dioxide concentrations for the last 350,000 years. These combined proxy sources suggest that for the past million years, glacial advances have lasted around 80,000 years and warmer periods (like our current interglacial period) have lasted about 20,000 years (figure 10.1). These cyclical changes can be attributed to Earth's orbital and tilt cycles and the resulting changes in solar radiation. The beginning and end of the interglacial warm periods are rather abrupt; global temperatures change by 20°F within several hundred years. The last glacial advance, or glaciation, ended 12,000 years ago and the climate has been generally warming since then, with significant fluctuations of both negative and positive warming.

Today's Climate Changes

For about the last hundred years, scientists have made enough weather observations and understand the climate system well enough to agree unanimously that climates vary from year to year and exhibit longer-term trends over time. There is also a consensus that in the last 120 years, average official measurements for all parts of the earth where data are available show that the earth is warming. It is also largely agreed that the observed warming has been more pronounced in the last 30 to 40 years than was the case earlier in the instrumental record. What changes could have caused this?

The world population has doubled from 3.5 billion people to over 7 billion people. This has happened in just 40 years (figure 10.2). Around the globe, we are increasingly reliant on energy for our livelihoods, our homes, and for transportation. The production and consumption of this energy and the changes to the land surface produced by human activities contribute either directly or indirectly to the heating of our planet. One way is through the release of carbon dioxide and other gases. These gases absorb energy radiated up from the earth's surface and re-radiate much of it back down to the surface. This process, known as the greenhouse effect, is completely natural, but is intensified by human activities.

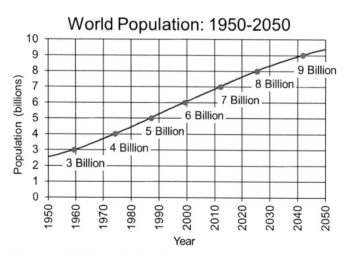

Figure 10.2. Global population change. Source: U.S. Census Bureau (2016).

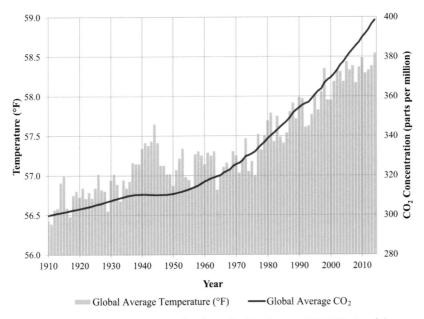

Figure 10.3. Global temperature and carbon dioxide. Source: NOAA National Centers for Environmental Information (n.d.a).

The growing global population and the increased use of carbon-based fuels have increased carbon dioxide levels from around 300 parts per million (ppm) around the year 1900 to over 400 ppm today (NOAA Earth System Research Laboratory 2016). Figure 10.3 shows the accelerated increase in carbon dioxide concentrations and resulting global temperature increases since about 1950. The temperature increases the earth is experiencing cannot be explained by natural cycles.

Long-term temperature-monitoring sites in some locations have become affected by encroaching urbanization. Vehicles and air conditioners emit heat and brick and asphalt store heat. This creates a likely increase in temperature due to changes in local conditions. But moving a temperature-recording station away from urban influences also means restarting the record. Thus, the long-term observations continue to record not only the local representation of the regional or global rise in temperature but also the localized influences of urbanization. When only the observations from non-urbanized U.S. locations are considered,

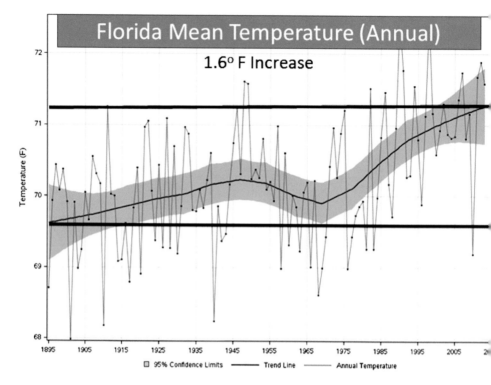

Figure 10.4. Statewide mean temperature in Florida, 1895–2015 Source: NOAA National Climatic Data Center (n.d).

the temperature has increased at the rate of approximately 1.6°F over the past century (Watts 2009).

The U.S. average temperature mimics the global trend. In 2016, the average temperature across global land and ocean surfaces was 1.69°F above the twentieth-century average and the hottest year in the 137-year record (NOAA National Centers for Environmental Information 2017). Near cities, the temperature increases have been much higher. Most of this increase has occurred since about 1960. Even though this mean national trend is not representative of the local patterns everywhere in the United States, the pattern of mean temperature for Florida closely mimics this national trend (figure 10.4).

Global and Local Changes in Temperature

Despite the fact that global average temperatures are warming, some regions have experienced colder-than-normal to record cold temperatures at times. One way to understand how localized cooling within a region can occur during global warming is to realize that in the winter, the poles have little sunlight and still get very cold. Although heat is transported from the tropics and subtropics to the middle and polar latitudes, the exchange takes place both ways. The air currents that flow poleward carrying warm air masses eventually flow back toward the equator carrying cold air masses. These cold air masses can initiate winter storms with blizzards and frigid temperatures. This can be compared to opening a refrigerator door inside a warm house. The "local" person standing in front of the open refrigerator door may be cooled, although the net effect of any refrigerator, open or closed, is to heat the house. This happens because the cool air inside the refrigerator is created by pumping the heat from inside the refrigerator to the outside. The earth is experiencing more warm temperature extremes than cold temperature extremes, and the number of record high-temperature days is increasing and the number of record cold-temperature days is decreasing.

Precipitation patterns have changed too. An analysis of weather observations shows that daily and five-day precipitation intensities have generally increased in the past 30 years, but so have periods of drought; the number of years with anomalously high or anomalously low precipitation has increased (Melillo, Richmond, and Yohe 2014).

The Little Ice Age

Some suggest that much of the observed warming is to be expected because the beginning of the period of observation using instruments coincides with the end of a naturally cold period, the Little Ice Age. The implication is that temperatures are increasing over time naturally, but this assumption doesn't account for influences from the record amount of greenhouse gases, most notably carbon dioxide, that has been added to the atmosphere by human activities.

The term "ice age" is used to describe the period from about AD 1450 to AD 1850—the Little Ice Age. It should be noted that even though the term Little Ice Age is well known, it is misleading. Real ice ages last for millions of years. The frigid climate and scarce resources during the Little Ice Age contributed to the harsh circumstances that prevailed during the Thirty Years' War in Europe (1618–1648) and the brutal winter at Valley Forge in 1777–1778. Coupled with the Little Ice Age, sulfate aerosols emitted to the atmosphere by the eruption of Mount Tambora in Indonesia in 1815 reflected, scattered, and absorbed enough additional sunlight across the earth to cause the "year without a summer" in the Northern Hemisphere in 1816. Volcanic eruptions, a decrease in solar output, and other explanations have been proposed as causes of the Little Ice Age.

Sea Level Rise

Global sea levels have risen about eight inches over the past century. Continued global temperature increases are expected to accelerate the rate of sea level rise. The rate of local sea level change at various locations in Florida differs from the global average because of geological nuances that may cause local land to sink (as we see in New Orleans), a process that exacerbates the effects of global sea level rise. In other places, such as parts of Alaska and Scandinavia, the land is rising relative to sea level because it is still "rebounding" after the tremendous weight of ice sheets which have melted.

Changes in global sea level due to a change in water volume is known as eustatic change, and the local- to regional-scale sea level change related to locally changing heights of land, such as in New Orleans, Alaska, and Scandinavia, is known as isostatic change. Floridians should be interested in understanding as much as possible about the possible range of changes in global sea level for obvious reasons. In Florida, 14 tidal gauges measured an average rate of sea level rise at 8.26 inches per century. The NOAA Key West tide gauge (figure 10.5) indicated a rise in sea level of 8.76 inches during the past century (NOAA 2013a). Short-term weather patterns have an impact on coastal water levels, as we see at Key West, which has experienced short-term rises and falls of sea level. But

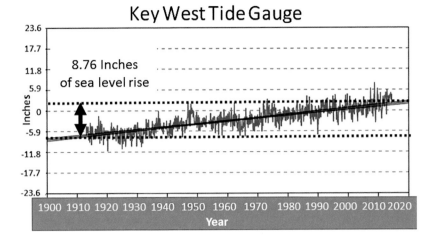

Figure 10.5. Water level measurements from the Key West tidal gauge indicating a rise of 8.76 inches. Source: NOAA (2013a).

the tidal-gauge record shows an overall rise in sea level over the past century.

Future Temperature Scenarios

Scientists who contributed to the 2014 U.S. Global Change Research Program National Climate Assessment Report (Melillo, Richmond, and Yohe 2014) developed predictions based on climate models using two scenarios: one in which fossil fuel emissions continue to increase and one in which fossil fuel emissions are reduced substantially. Temperatures in the two scenarios are forecasted to be similar until about 2050 but to diverge substantially after that time. By 2100, the amount of warming in Florida is likely to be alarming—in the range of 3°F for the reduced-emissions scenario and up to 6°F for the increased-emissions scenario. The reduced-emissions model predicts another 1°F of warming for the inland sections of the Southeast region of the United States; the increased-emissions model predicts an additional increase of 2°F for that region of the country.

Scientists predict that as average temperatures in Florida rise under global warming scenarios, the frequency of extreme warm events will also rise. While nearly all of Florida experienced an average of fewer than 15 days per year with a maximum temperature above 95°F for the 1971–2000 period, model projections suggest that the 2041–2070 period will include an average of over 75 such days per year for nearly all of Florida; the reduced-emissions scenario predicts slightly fewer such days. Similar warming in both scenarios is forecasted for minimum temperatures. NOAA's National Centers for Environmental Information and Cooperative Institute for Climate and Satellites note that from 1971 to 2000 the Florida Panhandle experienced an average of 20–40 subfreezing nights per year; the area south and east of Lake City experienced fewer than 20. In the increased emissions scenario, all of Florida is forecasted to experience fewer than 20 freezing nights per year in the 2041–2070 averaging period; the reduced-emissions scenario predicts a slightly less substantial decrease (Kunkel et al. 2013).

Future Sea Level Rise Scenarios in Florida

Florida's coastal terrain is nearly flat and is barely above sea level with little variation. Most of the state's 20 million residents are clustered near sea level along the coasts. In the short term, slowly rising sea levels inhibit coastal drainage. This creates problems, particularly during high tides, when heavy rains don't drain as fast, creating nuisance flooding. Larger problems occur when the intense winds of hurricanes and other coastal storms temporarily create more sudden and more drastic changes to coastal water levels than Florida has experienced in the past. During such extreme events, strong onshore winds push coastal waters shoreward, creating storm surge. The impact of storm surge is similar to that of a tsunami and can be devastating to coastal communities. Storm surge is highest where the shallow coastal waters extend far offshore, like those found along the gulf coast of Florida. More rapid changes in sea level rise create an adaptation problem whereby residents in areas of vulnerable coastal development must either leave or modify their homes before the "big one" arrives, without having the luxury of knowing the deadline.

The forecasts for future sea level rise depend on global temperatures and the amount of emissions humans produce on the planet. Scenarios of global sea level rise from the U.S. National Climate Assessment indicate that at the lowest end (the most conservative estimate, one that is not likely) the trend from the past century will continue. At the highest end (the least conservative estimate, which is also not likely), the scenario predicts a devastating six and a half feet of sea level rise. Nobody can know for sure what the actual rate will be, but the range between the best case and the worst case is most likely. Even though that window has a wide range of uncertainty, it is obvious that with Florida's huge expanse of coast, the exposure to this hazard is enormous. Rapidly increasing water levels would mean that people from coastal areas would need to be relocated to higher ground at immense costs. As the warming continues and oceans get deeper due to sea level rise, we will likely also see changes in ocean currents and a change in weather patterns.

Water Woes

Florida ranks first among the states in percentage of the population over age 65 (17.1 percent) (U.S. Census Bureau 2015). As a result, the impacts of climate change on people are likely to be more significant in Florida than elsewhere. The high population density (Florida ranks eighth among the states) exerts additional pressure on natural resources that are affected by global climate change, most notably freshwater availability. Some areas could experience a more rapid onset and increased severity of droughts.

Florida's natural environment also makes the global climate change discussion particularly important, especially as it relates to the increasing severity of coastal storms and increasing sea level. According to the NOAA's National Coastal Population Report (2013b), although only 39 percent of the nation's population lives in coastal shoreline counties, nearly 79 percent of Florida's population lives in such counties. In addition to the loss of coastal land, rising sea levels pose a significant threat to Florida's livelihood in other related ways. Rising sea levels create saltwater intrusion into freshwater areas (a process that would be aided by Florida's many porous limestone aquifers), leading to reduced freshwater availability and wetland destruction. On the positive side, at least one study shows that unlike most of the rest of the southeastern United

States, most of Florida is expected to experience an increase in water yield for 2010–2060. Of course, the uncertainty of such predictions in the context of an ever-increasing population invites a cautious approach to estimating potential water availability and planning for water use.

Threats to Flora and Fauna

When waters are warmer, algal blooms increase exponentially. When these blooms die off, the oxygen availability in aquatic ecosystems drops quickly, creating hypoxia (lack of oxygen) as the plant materials decompose. Hypoxia leads to fish kills. As waters rise and change the landscape, ecosystems become vulnerable and susceptible to new influences of inundation. As shorelines are pushed inland, plant and animal species will need to migrate, but this might not be possible when manmade structures such as sea walls impede their progress. Other effects of this warming could be detrimental to the world as we know it. When climate changes have a negative impact on local species and shift bioclimatic and epidemiological zones, the door is open for an increase in the destructive effects of invasive species.

Invasive species are already a massive problem in Florida. According to the Florida Fish and Wildlife Conservation Commission (2017), over 500 fish and wildlife nonnative exotic species have found homes in Florida. This statistic does not account for all of the plant species. When nonnative species present a threat to native species or cause economic damage they are deemed to be invasive. In Florida, invasive animal species include aggressive Asian tiger mosquitoes that bite during the daytime and have the ability to vector many viruses, such as West Nile virus; the carnivorous Nile monitor lizard, which can grow to five feet in length; the large and toxic cane toad; the Asian walking catfish; and the deadly giant Burmese python. The more than 100 species of plants that are deemed invasive in Florida include melaleuca; Brazilian pepper; Australian pine, which crowd out other vegetation; Old World climbing fern and Kudzu vines, both of which choke out native trees; and the notorious water hyacinth that blocks inland waterways.

Florida could be impacted by increased insect infestations and the expansion of tropical diseases such as malaria, dengue, and the Zika virus, which causes birth defects. Other impacts might include the more destructive influence of allergens and the respiratory impacts of higher concentrations of surface-level ozone, which occurs most often during the spring months. The impacts on human health and wellness will hit Florida disproportionately hard because of its large and impoverished elderly population.

Future Tropical Cyclones

Warmer oceans could potentially increase energy to fuel tropical cyclones. The scientific study of hurricanes and other storm activity during past climate regimes, known as paleotempestology, indicates that as ocean waters warm, more abundant and stronger storms tend to occur. Paleotempestology is a relatively new scientific discipline that examines evidence of storms that happened long ago. One method of studying past hurricanes is to take core samples from sediment ponds adjacent to beach sand dunes. During hurricanes, storm surges overwash beach sand dunes and deposit a layer of light-colored sand across dark mucky sedimentary back-bay ponds. These sandy layers in sediment cores of muck tell the story of storms from prehistory. One such study in southwest Florida by Ercolani and his team (2015) showed that there were fewer storms in the past 500 years, which included the Little Ice Age, than there were in the previous 500 years, from around AD 1000 to the 1500s when Northern Hemisphere temperatures were warmer. The study also revealed an uptick in the number of storm surge events in recent years. Under scenarios of continued global warming and continued rising sea levels, Florida is likely to pay a higher price than most. As water levels rise, homes, roads, and businesses will be gobbled up by the sea, forcing great change along the coast.

Benefits of a Warming Climate

Many have noted that warming is likely to provide some benefits. For example, increased growing season length may allow for higher agricultural and forest yields. However, in Florida, where the growing season is already near year-round, little would be gained. Instead, much could be lost, such as an increase in the number and incidences of tropical disease and an increase in levels of weed or pest infestation in agriculture. Others have noted that warming may allow for net reductions in personal energy consumption. But once again, this is unlikely to be true in Florida, where savings due to the warming of the few cold days throughout the year will not come close to offsetting the increased spending on cooling during the many hot days. Others mention that warming may bring cost savings in the transportation sector because there will be less need for snow removal and less likelihood of closed or dangerous road networks. Florida will likely not benefit since problems with cold weather are minimal there at present, though a reduction in the number of annual freezes would likely provide economic benefits to the agriculture industry. And finally, lifestyle benefits such as the likely reduced number of days of closed schools and businesses due to inclement weather and additional opportunities for outdoor recreation are sometimes mentioned as a side effect of warming. But once again, neither of these benefits would be felt in Florida, as few businesses close for inclement weather (except for tropical cyclones), and outdoor recreation persists year-round already.

The Future Is in Our Hands

So what is the concern? Climate change will bring many outcomes. As humans contribute to a warmer earth, the rapid changes in climate may have a catastrophic effect on life. Life forms that cannot adapt to climatic changes may perish. As sea levels rise, human habitations in low-lying coastal areas will become inundated, resulting in mass migrations to higher ground. It is to be hoped that the inland migration will be to newer urban communities with more efficient designs. These new communities should provide shopping and activities within walking distance, resulting in healthier Floridians. Increased pedestrian traffic and an increased use of mass transit instead of using single-passenger automobiles would reduce traffic congestion and air quality issues. Fossil fuels are nonrenewable resources even in this age of expanded exploration technologies. Even if time eventually proves that the consequences of our activities are less severe than anticipated, society will benefit from less reliance on activities that depend on fossil fuels.

Glossary

adiabatic: Warming of air by expansion and cooling of air by compression in a process that does not involve any temperature changes due to external forces such as changes in solar radiation or changes in the wind direction/circulation.

advection: The lateral or horizontal movement of air of a different temperature or moisture content.

advection fog: Fog caused by air being moved horizontally over a colder surface, such as a snow-covered landscape or a cold ocean current, and being chilled from underneath.

aerosol: A suspended tiny solid particle above the earth's surface, also known as a particulate.

air-mass thunderstorm: A thunderstorm that initiates from a single cumulonimbus cloud without contributions from a more organized system of clouds within the storm.

anticyclone: Any enclosed area of high pressure, either quasi-stationary or migratory.

aquifer: A geologic structure, such as much of the karst landscape of Florida, that allows groundwater to pass through and accumulate.

Atlantic Multidecadal Oscillation: A cycle of warmer and then colder-than-normal water temperatures in the North Atlantic Ocean that lasts 20 to 40 years.

back-door front: A front that occurs when the steering flow pushes a cooler air mass toward the southwest instead of to the southeast or east, as is usually the case.

Bermuda-Azores high: A semi-permanent subtropical anticyclone over the North Atlantic Ocean that is located east or northeast of Florida.

bioswale: A vegetated low-lying land area that slows and retains rainfall runoff and removes silt and pollutants before the water flows into downstream areas.

bounded weak echo region: An area on a radar image that indicates vigorous updrafts associated with severe weather.

bow echo: A persistent boomerang-shaped storm signature on radar that is often indicative of widespread strong winds and may also be called a derecho.

climate: The long-term averages, extremes, and variations in weather conditions of a location.

cyclogenesis: The formation stage of a cyclone.

cold front: An elongated area at which a colder air mass is displacing a warmer air mass and bringing lower temperatures to the locations through which it passes.

condensation: The transformation of water from gaseous form (vapor) to liquid form in a process that releases latent energy.

convection: Any vertical movement in a fluid, such as the motion in turbulent air that allows for the vertical growth of clouds.

Coriolis force: A force that to observers of a rotating body appears to deflect objects in motion. On Earth, the apparent deflection is clockwise in the Northern Hemisphere and counterclockwise in the Southern Hemisphere.

counter-trade winds: A belt of winds that blows from southwest to northeast in the Northern Hemisphere and northwest to southeast in the Southern Hemisphere that is caused by the deflection of trade winds to the right in the Northern Hemisphere and to the left in the Southern Hemisphere once the air crosses the equator.

cumuliform cloud: A convective cloud that is typically taller than it is wide and contains water near its base, and if tall enough, ice near its top; often the result of free convection in the presence of abundant moisture.

cumulonimbus cloud: A tall, well-developed cumuliform cloud, often with an anvil-shaped top that is producing lightning.

deepening: The intensification of a cyclone.

deposition: A change in the state of water from a vapor (gas) directly to a solid (ice), bypassing the liquid phase and releasing latent energy in the process.

derecho: A persistent arc-shaped, thunderstorm complex that creates widespread destructive straight-line wind and is sometimes called a bow echo.

dewpoint temperature: The temperature to which the air would need to cool in order to become saturated. As dewpoint temperature approaches the air temperature, the air becomes closer to saturation. Meteorologists use this measure to depict the amount of moisture in the atmosphere.

Dixie Alley: The zone where tornadoes occur frequently extending from Texas through the U.S. Southeast, including northern Florida.

double sea breeze: An effect in coastal circulation over peninsulas in which sea breezes form on each coast and propagate inland, creating a convergent collision that accelerates rising motion and creates greater precipitation.

downburst: The sudden downrush of air from a collapsing or surging thunderstorm.

dry adiabatic lapse rate: The rate at which temperature either decreases because a mass of unsaturated air expands as it rises or increases because a mass of unsaturated air compresses as it sinks. Meteorologists use this measure to assess the likelihood of rising motion that can cause clouds and precipitation to form.

dryline: The boundary between moist air and much drier air in the absence of a strong temperature gradient.

eccentricity: The degree to which Earth's orbit deviates from a perfect circle at a given point in geologic time.

Ekman flow: The turning of the ocean currents in response to the rotation of the earth such that the net transport of upper layers is at a 90-degree angle to the right of the surface wind flow in the Northern Hemisphere or at a 90-degree angle to the left of the surface wind flow in the Southern Hemisphere.

El Niño: A phenomenon that occurs every few years in which the normally cold ocean waters in the tropical Pacific Ocean adjacent to Peru become warmer, the northeast and southeast trade winds over the tropical Pacific Ocean weaken, and colder-than-normal waters persist in the western tropical Pacific Ocean near Indonesia. This combination disrupts normal atmospheric circulation patterns and weather conditions far beyond the Pacific Ocean, including Florida.

Enhanced Fujita scale: A system for categorizing the intensity of tornadoes that meteorologists began to use in 2007.

eustatic: An effect that is worldwide, as in sea level changes.

evaporation: The transformation of water from liquid form into gaseous form (vapor) in a process that absorbs latent energy.

evaporation fog: Fog that forms when water evaporates into air, saturating it and causing it to condense; includes precipitation fog and steam fog.

eye: The central part of a tropical cyclone.

eyewall: The part of a tropical cyclone closest to the eye that generally has the strongest winds.

eyewall replacement: A process that occurs in some well-developed hurricanes, typhoons, and cyclones that involves the dissipation of the eyewall and replacement with a new one, often larger than the first but sometimes accompanied by weakening winds.

Ferrel cell: The circulation feature between the polar cell and the Hadley cell in the Northern and Southern Hemispheres that includes the mid-latitude westerly winds.

Florida Current: The ocean current formed as the Loop Current exits the Gulf of Mexico through the deep water of the Florida Straits between Florida and Cuba.

fog: A cloud near the ground.

forced convection: Rising motion initiated by a mechanism such as a front or increased topography.

free convection: A phenomenon in the atmosphere in which a mass of air that is warmer than the surrounding air begins to rise spontaneously, without any mechanisms such as fronts that might force it to rise.

freezing rain: A form of winter precipitation that occurs as snow forms in a subfreezing layer of air aloft, melts on descent through a warmer layer, and then refreezes upon contact with a subfreezing surface such as a windshield, road, or bridge.

frontogenesis: The development of a front, particularly a cold front.

Fujita scale: A system for categorizing the intensity of tornadoes that was used until 2007.

funnel cloud: A downward protrusion from a cloud base that supports rotating, tornadic motion that may or may not touch down to the surface as a tornado or a waterspout.

greenhouse effect: A natural phenomenon in which carbon dioxide and other gases absorb radiant energy emitted by the earth's surface and reradiate much of it back down to the surface, warming the earth's lower atmosphere. The rate at which this occurs is affected by human activities.

ground fog: A type of radiation fog characterized by development very low to the ground that is caused by very humid conditions near the surface.

Gulf Stream: The great warm ocean current that brings tropical water from the Caribbean northward just east of Florida and flows toward northern Europe.

gustnado: A short-lived whirlwind caused by a downburst of air from a thunderstorm; lasts from seconds to a couple of minutes.

gyre: A large circular ocean surface current associated with large semi-stationary areas of high pressure.

Hadley cell: The vertical circulation feature that includes the trade winds and the ITCZ and extends from the equator to around 30 degrees in the Northern and Southern Hemispheres.

humid subtropical climate: A type of climate characterized by abundant precipitation year-round and mild winters; the temperature of the coldest winter month may dip below 64°F. This is the climate of most of Florida.

hurricane: The name given to the most intense category of tropical cyclone in the Atlantic and eastern Pacific basins.

hypoxia: A general lack of oxygen in an aquatic system.

ice age: Any time in Earth's history when there is "permanent" ice somewhere for many years through summer and winter.

intertropical convergence zone (ITCZ): The belt around the earth near the equator that oscillates seasonally and is characterized by persistent low atmospheric pressure and frequent cloud cover.

isostatic: Elevation changes that are local to regional in scale; the opposite of eustatic.

isotherm: A line of equal temperature on a weather or climate map.

karst landscape: A type of rock structure dominated by limestone in which calcium carbonate has become dissolved by slightly acidic rainwater; associated with sinkholes, underground drainage, and caverns.

lake breeze: A surface wind that blows from a lake onshore and typically occurs during the afternoon; local to regional in scale.

lake-breeze front: The leading edge of relatively cool air that blows onshore from a lake.

land breeze: A local or regional surface wind that blows from inland to a sea, lake, or ocean and typically occurs during the nighttime into early morning.

land-breeze front: The leading edge of air that blows offshore from inland associated with a land breeze.

La Niña: A phenomenon that occurs every few years in which the normally cold ocean waters in the tropical Pacific Ocean adjacent to Peru become even colder than normal, the northeast and southeast trade winds over the tropical Pacific Ocean strengthen, and warmer-than-normal waters persist in the western tropical Pacific Ocean near Indonesia. This combination disrupts normal atmospheric circulation patterns and weather conditions far beyond the Pacific Ocean, including Florida.

latent energy or latent heat: Energy that is released or absorbed as water molecules change phase.

latitude: An imaginary line on the geographic grid that runs west to east and measures distances north or south of the equator, ranging from 0° at the equator to 90° at the North Pole and South Pole.

Little Ice Age: The period from about A.D. 1450–1850 that is characterized by colder conditions globally than we experience today.

longitude: An imaginary line on the geographic grid that runs north to south and measures distances east or west of the prime meridian, ranging from 0° at the prime meridian to 180° in the Pacific Ocean.

longwave radiation: Electromagnetic radiation characterized by wavelengths greater than 4.5 millionths of a meter; almost entirely composed of energy emitted by the earth and its atmosphere.

Loop Current: The semi-permanent circulation feature in the eastern Gulf of Mexico that moves water from the Caribbean Sea north through the Yucatán Channel and then back south and into the Florida Current and eventually the Gulf Stream.

meteotsunami: A dangerous ocean wave produced by fast moving sudden and substantial changes in atmospheric pressure over continental shelf waters.

Milankovitch cycle: Any of the three orbital (i.e., eccentricity) or axial (i.e., tilt and wobble) variations that are believed to cause Earth's climate to change on time scales of thousands of years.

millibar (mb): A unit of pressure equivalent to 0.0145038 pounds per square inch.

mixed layer: The vertical layer from the surface of the earth to a height in the lower troposphere where abundant vertical mixing and turbulence occurs.

mixing ratio: The mass of water molecules divided by the mass of all other molecules in the atmosphere; an indicator of atmospheric humidity.

moist adiabatic lapse rate: The rate at which temperature decreases due to expansion as a mass of saturated air rises; used by meteorologists to assess the likelihood of rising motion that can cause clouds and precipitation to form.

northeast trade winds: A belt of winds that flows from northeast to southwest between the Intertropical Convergence Zone and the subtropical highs of the Northern Hemisphere.

occluded front: A cold front that has overtaken a warm front as it is pushed in a counterclockwise direction around a cyclone. The colder air behind the cold front then interfaces at the surface of the earth with the cooler air ahead of the warm front, and the warmer air that had been between the cold front and warm front is lifted above the surface.

orographic lifting: A mechanism of forced convection that is initiated when moving air encounters rugged topography.

outflow boundary: The lateral rush of surface air pushing out ahead of a thunderstorm.

Pacific-North America (PNA) pattern: A representation of the amplitude and position of the 500-mb ridge-and-trough configuration over North America at a given time. A high-index PNA pattern represents an amplified ridge over western North America and a trough over eastern North America; a low-index PNA pattern represents a de-amplified ridge over western North America and a trough over eastern North America.

paleotempestology: The scientific study of ancient storms by examining proxy evidence left by those storms.

particulate: A tiny solid particle suspended above the earth's surface; also known as an aerosol.

Polar cell: One of two circulation cells (one in the Northern Hemisphere and the other in the Southern Hemisphere) characterized by sinking air near the pole, polar easterlies with some equatorward flow adjacent to the sinking air, rising motion near 60° of latitude, west-to-east flow, and some poleward flow aloft.

polar easterlies: The system of east-to-west blowing winds near the surface between the pole and the Ferrel cell in the Northern and Southern Hemispheres.

polar jet stream: A large, meandering, west-to-east river of air in the upper troposphere, near the boundary between the Polar cell and the Ferrel cell.

precipitation fog: A type of evaporation fog that forms when rain falls into cooler and slightly drier air and begins to evaporate.

pressure: Force per unit area, such as the force exerted by a column of air on the surface of the earth.

prime meridian: The 0° line of longitude that runs through Greenwich, U.K.

radiation fog: Fog that develops as the combination of nighttime longwave radiational cooling under clear skies and nearly saturated conditions produces condensation near the surface.

radiosonde: An instrument transported via weather balloon that telemetrically reports atmospheric variables through a vertical slice of the atmosphere.

retrogression: Unusual east-to-west movement of a ridge and/or trough axis.

ridge: An elongated area of high atmospheric pressure.

rip current: A narrow, fast-moving channel of water that rushes out to sea from a beach and can be extremely dangerous to swimmers.

Saffir-Simpson scale: A categorization system from 1 to 5 that is based on hurricane wind speeds and indicates potential damage to structures.

savanna: A tropical or subtropical grassland with few trees.

sea breeze: A surface wind on the local to regional scale that blows from a sea or ocean onshore and typically occurs during the afternoon.

sea-breeze front: The leading edge of air that blows onshore from the sea or ocean.

sea fog: A type of advection fog that forms when warm moist air moves over colder ocean waters and cools the air to the dewpoint thereby creating fog.

shortwave radiation: Electromagnetic energy that is characterized by wavelengths less than 4.5 millionths of a meter; almost entirely composed of energy emitted by the sun.

Skew-T diagram: A tool meteorologists use that depicts the vertical profile of the temperature, humidity, wind speed, and temperature change that is expected as the result of rising or sinking parcels of air at a given place and time; includes data measured by radiosonde and reported telemetrically to a central receiving computer.

sleet: A form of winter precipitation that occurs as snow forms in a subfreezing layer of air aloft, melts on descent through a warmer layer, and then refreezes in a subfreezing air layer near the surface before hitting the ground.

southeast trade winds: A belt of winds that flows from southeast to northwest between the Intertropical Convergence Zone and the subtropical highs of the Southern Hemisphere.

squall line: A narrow line of violent thunderstorms moving ahead of a cold front, often accompanied by tornadoes and other hazardous weather.

stationary front: A cold front, warm front, or occluded front that becomes temporarily stalled.

steam fog: A type of evaporation fog that forms as cooler air overlies a warm inland water body and rising moist air from near the water body evaporates into the cooler air and saturates it.

stepped leader: A downward flow of negatively charged electrical energy from a cloud that meets an upward flow of positively charged electrical particles from the ground, setting the stage for a lightning strike.

storm surge: The accumulation of ocean water onshore, typically associated with hurricanes caused by windblown water over shallow continental shelf areas.

stratosphere: the layer of the atmosphere just above the troposphere characterized by isothermal temperatures and temperature inversions.

subsidence inversion: The process by which air aloft is warmed adiabatically through sinking that it becomes warmer than the air below it, which was not compressed as much.

subtropical anticyclone: A quasi-stationary, semi-permanent enclosed area of high pressure that exists just north of the trade winds in the Northern Hemisphere and just south of the trade winds in the Southern Hemisphere (but not all the way to the pole) and provides generally fair weather. In the Northern Hemisphere, subtropical anticyclones provide outward and clockwise circulation; in the Southern Hemisphere, they generate counterclockwise and outward flow.

subtropical jet stream: A large, meandering, west-to-east river of air in the upper troposphere near the boundary between the Hadley cell and the Ferrel cell.

supercell: A single, large, long-lived, rotating thunderstorm that often spawns one or more tornadoes.

tornado: A violent, rotating, vertical column of air that emerges from a cumulonimbus thunderstorm cloud and causes very strong winds at the earth's surface.

Tornado Alley: The part of the United States from Texas north to Iowa where tornadoes occur more intensely and more frequently than virtually anywhere else on earth.

trade winds: The global system of persistent near-surface winds from northeast to southwest in the Northern Hemisphere and from southeast to northwest in the Southern Hemisphere.

transpiration: The loss of water from the surface to the atmosphere by way of the roots, stems, and leaves of plants.

Tropic of Cancer: The parallel of latitude at 23.5°N; the farthest poleward latitude in the Northern Hemisphere that experiences the direct rays of the sun in the summer.

Tropic of Capricorn: The parallel of latitude at 23.5°S; the farthest poleward latitude in the Southern Hemisphere that experiences the direct rays of the sun in the summer.

tropical climate: A climate type, such as in southern Florida, in which the average temperature of the coldest month remains above 64°F and is not arid.

tropical cyclone: A generic term for low-pressure systems or storms, such as hurricanes, that form in or near the tropics.

tropical depression: A tropical disturbance that gains cyclonic circulation.

tropical disturbance: A discrete system of organized convection (showers or thunderstorms) that originates in the tropics or subtropics, does not migrate along a frontal boundary, and maintains its identity for 24 hours or more.

tropical storm: A tropical cyclone that attains wind speeds of at least 39 mph.

tropical wet-dry climate: A type of tropical climate in which precipitation is abundant in some months but absent in others; also known as a tropical savanna climate.

tropopause: The top of the troposphere.

troposphere: The lowest layer of the atmosphere characterized by the presence of nearly all of the atmosphere's water and therefore nearly all of the weather and climate.

trough: An elongated area of low atmospheric pressure.

tsunami: A destructive ocean wave produced by an earthquake or landslide that may occur thousands of miles away.

typhoon: The most intense category of tropical cyclone in the western Pacific basin.

ultraviolet radiation: High-intensity, shortwave electromagnetic radiation from the sun that is largely absorbed in the ozone layer, making the terrestrial earth habitable.

upwelling: The upward movement of cold ocean water to the surface.

urban heat island: An isolated area of higher temperatures created by heat retention within city structures.

valley fog: A special type of radiation fog that forms in locally low elevations as colder, denser air moves downward in response to gravity.

vorticity: The spinning motion of a circulation in the atmosphere.

warm front: The elongated boundary between a warmer air mass and a colder air mass where the warmer air mass displaces the colder air mass.

waterspout: A tornado-like vortex that forms over water.

westerlies: The generally west-to-east flow of air in the middle latitudes of both the Northern and Southern Hemispheres.

wind shear: A change in wind speed with height or distance.

References

Albrecht, R., S. Goodman, D. Buechler, R. Blakeslee, and H. Christian. 2016.
   Where are the lightning hotspots on Earth? *Bulletin of the American Meteorological Society* 97: 2051–2068. doi:10.1175/BAMS-D-14–00193.1.
Associated Press. 1962. Savage force of tornado seen. Stevens Point Daily Journal
   (WI), April 2, 1962. Reprinted at *GenDisasters.com*. Accessed November 20,
   2016. http://www.gendisasters.com/florida/11167/milton-fl-tornado-strikes-
   apr-1962.
Atlantic Oceanic and Meteorological Laboratory. 2016. Hurricane Research Division: Re-Analysis project, 1871. Accessed November 23, 2016. http://www.
   aoml.noaa.gov/hrd/hurdat/1871.htm.
Ballingrud, D. 2002. It could happen here: Hurricane Andrew 10 years later. *St.
   Petersburg Times* August 24. Accessed November 10, 2016 from http://www.
   sptimes.com/2002/webspecials02/andrew/day4/story1.shtml.
Blake, E. S., E. N. Rappaport, and C. W. Landsea. 2007. The deadliest, costliest,
   and most intense United States tropical cyclones from 1851 to 2006 (and other
   frequently requested hurricane facts). NOAA Technical Memorandum NWS
   TPC-5. Accessed February 6, 2-17. Accessed February 6, 2017. http://www.
   nhc.noaa.gov/pdf/NWS-TPC-5.pdf.
Centers for Disease Control. 1996. Deaths associated with Hurricanes Marilyn
   and Opal—United States, September–October 1995. *MMWR Weekly* 45(2):
   32–38. Accessed November 29, 2016. https://www.cdc.gov/mmwr/preview/
   mmwrhtml/00040000.htm.
Changnon, S. A., and T. R. Karl, 2003. Temporal and spatial variations of freezing
   rain in the contiguous United States: 1948–2000. *Journal of Applied Meteorology* 42(September): 1302–1315.
Collins, J. M., R. Ersing, and A. Polen. 2017. Report for the Natural Hazards
   Center Quick Response Grant: Evacuation Behavior Measured at Time of Expected Hurricane Landfall: An Assessment of the Effects of Social Networks.
Collins, J. M., and P. Flaherty. 2014. Keeping an "eye" on tropical research data:
   The NOAA hurricane hunters, their missions and their recent work with the

University of South Florida to archive historical information. *The Florida Geographer* 45: 14–27.

Collins J. M., C. H. Paxton, and A. N. Williams. 2009. Precursors to southwest Florida warm season tornado development. *Electronic Journal of Operational Meteorology* EJ12. Accessed November 10, 2016. http://nwafiles.nwas.org/ej/pdf/2009-EJ12.pdf.

Collins, J. M., A. N. Williams, C. H. Paxton, R. J. Davis, and N. M. Petro. 2009. Geographical, meteorological, and climatological conditions surrounding the 2008 Interstate-4 disaster in Florida. In *Papers of the Applied Geography Conferences*, ed. Lisa M. B. Harrington and John A. Harrington, 153–162. Binghamton, NY: Applied Geography Conferences.

Collins, W. G., Paxton, C. H., and Golden, J. H. 2000. The 12 July 1995 Pinellas County, Florida, tornado/waterspout. *Weather and Forecasting* 15:122–134.

Davis, D., and D. Millott. 1966. Tornadoes kill 8; Damage in millions. *St. Petersburg Times*, April 5, 1966, 1A. Accessed November 21, 2016. https://news.google.com/newspapers?nid=feST4K8J0scC&dat=19660405&printsec=frontpage&hl=en.

Department of Commerce, NOAA, and Office of Marine and Aviation Operations. 2017. NOAA Hurricane Hunters. Accessed February 13, 2017. http://www.omao.noaa.gov/learn/aircraft-operations/about/hurricane-hunters.

Dolce, Chris. 2015. Florida is nearing 10 years with no hurricanes. *Hurricane News*, August 26. Accessed February 6, 2017. https://weather.com/storms/hurricane/news/florida-hurricane-drought-erika.

Ercolani, C., J. Muller, J. Collins, M. Saverese, and L. Squiccimara. 2015. Intense southwest Florida hurricane landfalls over the past 1,000 years. *Quaternary Science Reviews* 126(October 15): 17–25.

Florida Department of Agriculture and Consumer Services, Division of Forestry. 2010. *Wildfire risk reduction in Florida: Home, neighborhood, and community best practices.* Gainesville, FL: Sorter Printing. Accessed February 13, 2017. http://freshfromflorida.s3.amazonaws.com/Wildfire_Risk_Reduction_in_FL.pdf.

Florida Department of Environmental Protection. 2015. Aquifers. Accessed November 10, 2016. https://fldep.dep.state.fl.us/swapp/Aquifer.asp.

Florida Fish and Wildlife Conservation Commission. 2017. Florida's exotic fish and wildlife. Accessed February 13, 2017. http://myfwc.com/wildlifehabitats/nonnatives/.

Florida Forest Service. 2010. Wildfire risk reduction in Florida. Accessed November 10, 2016. http://freshfromflorida.s3.amazonaws.com/Wildfire_Risk_Reduction_in_FL.pdf.

Florida State Geological Survey. 1994. Florida's geological history and geological resources. FSGS Special Publication no. 35. Accessed October 5, 2016. http://

publicfiles.dep.state.fl.us/FGS/FGS_Publications/SP/SPPRIDE/SP35PRIDE/FSGS%20Special%20Publication%20No.%2035.pdf.

Frederick, E. 2009. *Ten seconds inside a tornado.* The Villages, FL: AM World Editions.

Grazulis, T. P. 1993. *Significant tornadoes, 1680–1991: A chronology and analysis of events.* St. Johnsbury, VT: Environmental Films.

Hagemeyer, B. C., and S. M. Spratt. 2002. Thirty years after Hurricane Agnes—The forgotten Florida tornado disaster. In *The 25th conference on hurricanes and tropical meteorology,* 422–423. San Diego, CA: American Meteorological Society.

Hurricane Research Division. 2014. Hurricane re-analysis project. Accessed February 6, 2017. http://www.aoml.noaa.gov/hrd/hurdat/Data_Storm.html.

Iowa State University of Science and Technology. 2017. Wind roses. Accessed February 7, 2017. http://mesonet.agron.iastate.edu/sites/windrose.phtml?network=WI_ASOS&station=EZS.

Keim, B. D., R. A. Muller, and G. W. Stone. 2007. Spatiotemporal patterns and return periods of tropical storm and hurricane strikes from Texas to Maine. *Journal of Climate* 20: 2498–3509.

Kunkel, K. E, L. E. Stevens, S. E. Stevens, L. Sun, E. Janssen, D. Wuebbles, C. E. Konrad II, C. M. Fuhrman, B. D. Keim, M. C. Kruk, A. Billet, H. Needham, M. Schafer, and J. G. Dobson. 2013. *Regional climate trends and scenarios for the U.S. national climate assessment. Part 2. Climate of the Southeast U.S.* NOAA Technical Report NESDIS 142-2. Washington, D.C. NOAA. Accessed February 7, 2017.https://www.sercc.com/NOAA_NESDIS_Tech_Report_Climate_of_the_Southeast_U.S.pdf.

Landsea, C. W., J. L. Franklin, C. J. McAdie, J. L. Beven II, J. M. Gross, B. R. Jarvinen, R. J. Pasch, E. N. Rappaport, J. P. Dunion, and P. P. Dodge. 2004. A reanalysis of Hurricane Andrew's intensity. *Bulletin of the American Meteorological Society* 85(November):1699–1712.

McDonald, W. F. 1935. The hurricane of August 31 to September 6, 1935. *Monthly Weather Review* 63(9): 269–271.

Melillo, Jerry M., Terese (T. C.) Richmond, and Gary W. Yohe, eds. 2014. *Climate change impacts in the United States: The third national climate assessment.* Washington, D.C.: U.S. Government Printing Office. Accessed February 6, 2017. http://s3.amazonaws.com/nca2014/low/NCA3_Full_Report_0a_Front_Matter_LowRes.pdf?download=1.

NASA (National Aeronautics and Space Administration). N.d. Worldview. Accessed February 13, 2017. https://worldview.earthdata.nasa.gov/.

———. 2001. Where lightning strikes. *NASA Science Beta,* December 5. Accessed October 16, 2016. https://science.nasa.gov/science-news/science-at-nasa/2001/ast05dec_1.

National Climatic Data Center. 1998. Historical station based wind climatology for the U.S. November. Accessed November 23, 2016. https://www.ncdc.noaa.gov/sites/default/files/attachments/wind1996.pdf.

National Hurricane Center. 2017. National Hurricane Center Tropical Cyclone Report: Hurricane Matthew (AL142016). Accessed April 13, 2017. http://www.nhc.noaa.gov/data/tcr/AL142016_Matthew.pdf.

National Severe Storms Laboratory. 2016. Warn on forecast fact sheet. Accessed August 13, 2016. http://www.nssl.noaa.gov/projects/wof/.

National Weather Service. 2013. Straight-line winds vs. tornado: What's the difference? *National Weather Service*, Northern Indiana Weather Forecast Office. Accessed February 27, 2017. https://www.weather.gov/iwx/2013_straight-line_winds_vs_tornado.

National Weather Service Forecast Office, Melbourne, Florida. 1998. The central Florida tornado outbreak of 22–23 February, 1998. Accessed November 23, 2016. www.weather.gov/media/mlb/surveys/mlbsep_outbreak98.pdf.

National Weather Service NCEP Central Operations. 2016. GFS North Polar Region: Available model cycles. Accessed February 13, 2017. http://mag.ncep.noaa.gov/model-guidance-model-parameter.php?group=Model%20Guidance&model=gfs&area=polar&cycle=20170209%2018%20UTC&param=500_vort_ht&fourpan=no&imageSize=&ps=model&fhr_mode=image.

National Weather Service Weather Prediction Center. 2009. Surface and upper air maps. Accessed February 13, 2017. http://www.spc.noaa.gov/obswx/maps/.

———. 2017. WPC's Surface Analysis Archive. Accessed February 13, 2017. http://www.wpc.ncep.noaa.gov/archives/web_pages/sfc/sfc_archive.php.

National Weather Service and NOAA. 1995. Hurricane Opal: October 4, 1995. Accessed February 13, 2017. https://www.weather.gov/mob/opal.

———. 2006. April 13, 2006: Southern Wisconsin hailstorm. Accessed February 13, 2017. https://www.weather.gov/mkx/041306_hail.

———. 2009. Weatherspout along the St. Johns River (6/26/09). Accessed September 21, 2009. http://www.srh.noaa.gov/jax/?n=weatherstory_september2009.

———. 2010. Record setting hail event in Vivian, South Dakota on July 23, 2010. Accessed February 13, 2017. https://www.weather.gov/abr/vivianhailstone.

Neely, Wayne. 2014. *The Great Okeechobee Hurricane of 1928: The story of the second deadliest hurricane in American history and the deadliest hurricane in Bahamian history*. Bloomington, IN: iUniverse.

NOAA (National Oceanic and Atmospheric Administration). N.d. Changing seasons. Accessed February 13, 2017. http://www.noaa.gov/resource-collections/changing-seasons.

———. 2005. NOAA mobilizes resources to aid in recovery from Hurricane Katrina. Accessed February 13, 2017. http://www.noaanews.noaa.gov/stories2005/s2494.htm.

———. 2013a. Mean sea level trend: 8724580 Key West, Florida. NOAA: Tides and currents. Accessed August 27, 2016 from https://tidesandcurrents.noaa. gov/sltrends/sltrends_station.shtml?stnid=8724580.

———. 2013b. National coastal population report: Population trends from 1970 to 2000. Accessed February 13, 2017. http://oceanservice.noaa.gov/facts/ coastal-population-report.pdf.

———. 2016. *Comparative climatic data for the United States through 2015.* Silver Spring, MD: NOAA; National Environmental Satellite, Data and Information Service; and National Centers for Environmental Information. Accessed December 19, 2016. from http://www1.ncdc.noaa.gov/pub/data/ccd-data/ CCD-2015.pdf.

NOAA Climate.gov. 2010. Tropical cyclone tracks. Accessed February 13, 2017. https://www.climate.gov/news-features/understanding-climate/tropical-cyclone-tracks.

NOAA Earth System Research Laboratory. N.d. NCEP/NCAR reanalysis project at the NOAA/ESRL Physical Sciences Division. Accessed February 27, 2017. https://www.esrl.noaa.gov/psd/data/reanalysis/reanalysis.shtml.

———. 2016. NOAA's Annual Greenhouse Gas Index. *NOAA Earth System Research Laboratory: Global Monitoring Division,* Accessed November 22, 2016. http://www.esrl.noaa.gov/gmd/aggi/.

NOAA Environmental Visualization Laboratory. 2014. Powerful storm brings threat of severe weather to East Coast, blizzard conditions to Midwest. Accessed February 13, 2017. https://www.nnvl.noaa.gov/MediaDetail2.php?MediaID =1508&MediaTypeID=1.

NOAA Geostationary Satellite Server. 2015. GEOS goes full disk. Accessed February 13, 2017. http://www.goes.noaa.gov/goesfull.html.

NOAA National Centers for Environmental Information. N.d.a. Global climate change indicators. Accessed February 13, 2017. https://www.ncdc.noaa.gov/ monitoring-references/faq/indicators.php.

———. N.d.b. Storm events database. Accessed November 23, 2016. https://www. ncdc.noaa.gov/stormevents/.

———. 2017. State of the Climate: Global Analysis, Annual 2016. Accessed August 23, 2016. http://www.ncdc.noaa.gov/sotc/global/201513.

NOAA National Climatic Data Center. N.d. State annual and seasonal time series. Accessed February 17, 2017. https://www.ncdc.noaa.gov/temp-and-precip/state-temps/.

———. 2008. Glacial-interglacial cycles. Accessed February 13, 2017. https:// www.ncdc.noaa.gov/paleo/abrupt/data2.html.

NOAA and National Weather Service. 1994. *Superstorm of March 1993.* Accessed February 13, 2017. www.weather.gov/media/publications/assessments/Super-storm_March-93.pdf.

NOAA and National Weather Service Storm Prediction Center. 2010. Observed sounding archive. Accessed February 13, 2017. http://www.spc.noaa.gov/exper/soundings/.

———. 2016a. Storm Prediction Center WCM page. Accessed February 5, 2017. http://www.spc.noaa.gov/wcm/.

———. 2016b. SVRGIS (updated: 14 March 2016). Accessed February 13, 2017. http://www.spc.noaa.gov/gis/svrgis/.

NOAA, National Weather Service, and Weather Forecast Office, Miami. 2010. Historic Cold Episode of January 2010. Accessed February 5, 2017. https://www.weather.gov/media/mfl/news/ColdEpisodeJan2010.pdf.

NOAAVisualizations. 2013. 20th Anniversary of the "Storm of the Century" March 1993. YouTube video. https://www.youtube.com/watch?v=tb5ElqmCdH0.

Pielke, R. A., R. L. Walko, L. Steyaert, P. L. Vidale, G. E. Liston, and W. A. Lyons. 1999. The influence of anthropogenic landscape changes on weather in south Florida. *Monthly Weather Review* 127(July): 1663–1673.

Ray, P. S., X. Du, and J. Rivard. 2014. Analysis of prospective systems for fog warnings. Unpublished report. FDOT contract number BDK83 997-19. Accessed November 10, 2016. http://www.fdot.gov/research/Completed_Proj/Summary_TE/FDOT-BDK83-977-19-rpt.pdf.

Schuyler, N., and J. Longman. 2010. *Not without hope*. New York: William Morrow Publishers.

Stewart, S. R. 2004. Tropical cyclone report: Hurricane Ivan, 2–24 September 2004. National Hurricane Center Report. Accessed November 25, 2016. http://www.nhc.noaa.gov/data/tcr/AL092004_Ivan.pdf.

U.S. Census Bureau. 2011. Historical census of housing tables. *United States Census Bureau*. Accessed February 13, 2017. https://www.census.gov/hhes/www/housing/census/historic/units.html.

———. 2015. *State & county quickfacts: Florida*. Accessed February 12, 2017. http://www.census.gov/quickfacts/table/PST045216/12.

———. 2016. International Data Base: World population, 1950–2050. Accessed February 13, 2017. https://www.census.gov/population/international/data/idb/worldpopgraph.php.

Watts, A. 2009. *Is the U.S. surface temperature record reliable?* Chicago: The Heartland Institute. Accessed August 4, 2016. https://wattsupwiththat.files.wordpress.com/2009/05/surfacestationsreport_spring09.pdf.

Webb, C. A., D. M. Bush, and R. S. Young. 1997. Property damage mitigation lessons from Hurricane Opal: The Florida Panhandle Coast. *Journal of Coastal Research* 13(1): 246–252.

World Heritage Encyclopedia. 2016. 1848 Tampa Bay hurricane. Accessed November 29, 2016. http://www.worldlibrary.org/articles/great_gale_of_1848.

Xian, Z. J., and R. A. Pielke. 1991. The effects of width of landmasses on the development of sea breezes. *Journal of Applied Meteorology* 30(9): 1280–1304.

Index

JENNIFER M. COLLINS is associate professor in the School of Geosciences at the University of South Florida.

ROBERT V. ROHLI is professor at Louisiana State University. He is the coauthor of *Louisiana Weather and Climate*.

CHARLES H. PAXTON is an American Meteorological Society certified consulting meteorologist.